Bamboos and Grasses

THE ROYAL HORTICULTURAL SOCIETY

Bamboos and Grasses

Jon Ardle

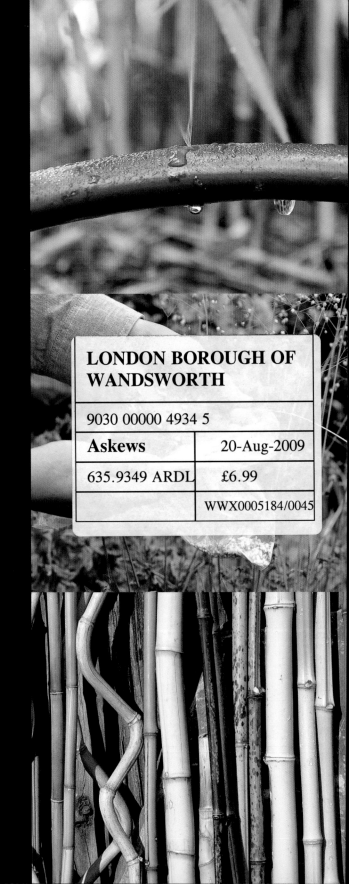

LONDON, NEW YORK, MUNICH,
MELBOURNE, DELHI

SENIOR EDITOR Zia Allaway
ACTING SENIOR DESIGNER Rachael Smith
MANAGING EDITOR Anna Kruger
MANAGING ART EDITOR Alison Donovan
DTP DESIGNER Louise Waller
PICTURE RESEARCH Lucy Claxton,
Richard Dabb, Mel Watson
PRODUCTION CONTROLLER Rebecca Short

PRODUCED FOR DORLING KINDERSLEY
Airedale Publishing Limited
CREATIVE DIRECTOR Ruth Prentice
DESIGNER Murdo Culver
EDITOR Helen Ridge
PRODUCTION MANAGER Amanda Jensen

PHOTOGRAPHY Sarah Cuttle, David Murphy

First published in Great Britain in 2007 by
Dorling Kindersley Ltd
Penguin Books Ltd
80 Strand
London WC2R 0RL

4 6 8 10 9 7 5 3

A CIP catalogue record for this book is available
from the British Library.

ISBN 9781405316835

Reproduced by Colourscan, Singapore
Printed and bound in Singapore by Star Standard

Discover more at
www.dk.com

Contents

Designing with grasses & bamboos

Whatever the setting, be it a gravel garden, Oriental-style design, or patio display, bamboos, grasses and grass-like plants can make it special. In this chapter, discover the beauty of these versatile plants, and how to create a range of exciting designs by combining them with flowers, shrubs and trees.

All-grass plantings

Planting a border with just grasses can provide year-round interest and be surprisingly colourful, with a sense of movement that shrubs or perennials simply cannot match.

Pictures clockwise from below

Colour contrast The combination of the relatively wide, bronze-red leaves of hook sedge (*Uncinia uncinata rubra*) and the thread-like, narrow green leaves of clump-forming *Festuca glauca* illustrates the contrast of hues that can be achieved using grasses. Both these plants prefer acid soils and are evergreen, requiring only an annual tidy in spring.

Mixing sizes and textures Some of the best large plants for the centrepiece of an all-grass planting are the many cultivars of pampas grass (*Cortaderia selloana*). Some can reach 3m (10ft) tall, with clumps of grey-green foliage and huge plumes of "ostrich feather" flowers in late summer (the dwarf cultivar 'Pumila' is better for small gardens). In the foreground are smaller *Stipa calamagrostis*, which flower for a long period, and one of the best blue-leaved grasses, *Panicum virgatum*.

Compact and modern Snow, or tussock, grasses (*Chionochloa*) are elegant, mounded plants with arching, plume-like flowers, like mini pampas grasses. This broad-leaved type, *C. flavescens*, is the hardiest tussock.

Winter structure Even deciduous grasses can contribute much to the winter garden. The bleached straw "bones" and flowerheads of *Miscanthus* can survive very harsh winters and look particularly attractive rimed with frost.

Grasses and bamboos in mixed borders

Grasses and bamboos can add an exotic touch to traditional herbaceous and mixed borders, whether sunny or shaded. Choose them carefully to integrate with the vigour and appearance of existing plantings for beautiful effects.

Pictures clockwise from top far left

Feathers and fire In this planting, grasses provide an effective foil for the vivid scarlet flowers of *Crocosmia*. The airy fawn flowerheads of Spanish oat grass (*Stipa gigantea*) add a hazy backdrop, throwing the flowers into even sharper relief, and harmonize with the softer flowers of feather grass (*Stipa tenuissima*). Both grasses will still have flowers long after those of *Crocosmia* have faded, creating a more subtle, but no less attractive, display.

Woodland walk Many bamboos are forest understorey plants in their native habitats and so are ideal for shaded, woodland schemes. Placing examples of the same species either side of a path not only contrasts with the shape of the trees forming the canopy, but leads the eye (and feet) further on. The upright bamboo stems and lower-growing rounded shrubs also create a harmonious picture.

Making a statement Bamboos are unsurpassed as large, bold, and exotic specimen plants, but for their impact to be undiminished, their companions must be chosen carefully. In the foreground of this planting, the yellow flowers of a spurge, *Euphorbia characias* subsp. *wulfenii*, echo the colour and form of the magnificent golden culms (stems) of a *Phyllostachys* bamboo, while the spidery white flowers of *Magnolia stellata* light up the background. Like the bamboo, the magnolia is a native of the Orient.

Moving screen In an exquisite show garden designed by Tom Stuart-Smith, the airy, translucent nature and sheer height of Spanish oat grass (*Stipa gigantea*) have been exploited to the full. The plants create a moving screen, waving above, but never obscuring, a naturalistic, pastel planting of salvias, marguerite daisies, stachys, and burgundy bearded iris.

Prairie planting

Inspired by the natural plant communities of the North American prairies and the Russian steppes, prairie planting is a popular, low-maintenance, and drought-tolerant style of gardening.

Pictures clockwise from right
Small scale Prairie plantings can be scaled down to suit smaller plots by using short grasses and perennials. Among the smaller grasses planted in this garden are *Molinia*, *Deschampsia*, and *Calamagrostis* x *acutiflora,* with Siberian iris (*Iris sibirica*), *Knautia macedonica,* daylilies (*Hemerocallis*), and *Salvia pratensis* among the perennials.
Late-season appeal One of the major attractions of prairie plantings is their long season of interest; in autumn, *Stipa gigantea* flowers are still going strong, as are the blue spires of *Perovskia atriplicifolia.* The spherical seedheads of globe thistles (*Echinops*) and sea hollies (*Echium*) help give the planting structure through winter.
Classic combination One of the most characteristic North American prairie plants is black-eyed Susan (*Rudbeckia*). With the hazy bronze flowers of *Stipa tenuissima*, a beautiful, long-lasting combination is created.

Meadow planting

Meadows are basically mixtures of grasses and wild flowers. Originally the result of traditional farming practices, meadows can be shrunk to fit a domestic garden.

Pictures clockwise from right

Perennial meadows Seed mixtures of perennial grasses and flowers are available to suit many soil types and situations. For relatively moist, semi-shaded positions, daisies, buttercups, and snake's-head fritillaries can be added to grassland as small "plug" plants or bulbs.

Annual mixes Probably the easiest type of meadow to establish is one based on a pre-formulated mix of annual grasses and wild flowers. Sown onto cleared ground, these give a beautiful flowering meadow of annual poppies (*Papaver*) and cornflowers (*Centaurea cyanus*) within weeks, but will need resowing every spring. Using native species makes meadows particularly valuable to wildlife.

Grass flowers The flowerheads of many grasses, such as fountain grass (*Pennisetum alopecuroides*, far right) and wild barley (*Hordeum jubatum*, below) are beautiful in their own right, particularly if grown in full sun when they take on tinges of pink. It is important to match the vigour of the grasses to those of the wild flowers: meadows actually succeed best on poor, dry soils, where it is easier for the flowers to compete with the grasses.

Gravel gardens

The graceful forms of grasses and bamboos look great when planted in gravel, where they usually grow well, relishing the sharp drainage. Place a weed-suppressing fabric under the gravel for a beautiful low-maintenance garden.

Pictures clockwise from left

Neutral backdrop Gravel helps to highlight individual grasses by providing them with a neutral backdrop. Clump-forming types are ideally suited to gravel gardens, but leave plenty of room between the plants or the effect is lost.

Golden grass *Stipa gigantea* 'Gold Fontaene', Spanish oat grass, makes a beautiful specimen for a gravel garden. It forms an upright mound around 75cm (30in) high, from which rise yellow flower stems in summer up to 1.8m (6ft) tall.

Bamboo foil Gravel sets off bamboos beautifully. Adding a few water-smoothed cobbles or pebbles enhances the look, their rounded edges contrasting with the straight lines of the bamboo canes.

Blue grass edging One of the most striking small grasses is *Festuca glauca* 'Elijah Blue', with its bright blue evergreen leaves. Forming tight, low clumps, it is excellent for edging paths or borders.

Contemporary and low-maintenance gardens

Given their diversity of colours, shapes, and textures, and the fact that most need very little attention, grasses and bamboos are ideal plants to accompany the sleek lines and cutting-edge materials of modern gardens. They are particularly suited to low-maintenance designs and "outdoor rooms".

Pictures clockwise from top left

Lawn alternative In a quirky take on the traditional lawn, this garden features a rectangle of low-growing green succulent sedum matting edged with a wide strip of arching, light-flowered feather grass (*Stipa tenuissima*). Unlike a lawn, neither requires mowing. Spires of white foxgloves push through the grasses, breaking up the expanse of green.

Unashamedly modern Forming the dramatic corner of an outdoor room, a galvanized square container houses a gorgeous *Phyllostachys bambusoides* 'Castillonii', with its orange, green-striped stems. The bamboo is underplanted with purple berberis and flanked by bold bearded iris, a purple-leaved hardy geranium, and, to chime with the bamboo, a switch grass (*Panicum*).

Mirror images The colour of versatile fescue, *Festuca glauca*, makes it ideal for use with modern materials, such as mirrored tiling, stainless steel, and black floor tiles. In this small, modern garden, a flowering blue *Festuca* in the centre of the display is flanked by another cultivar yet to bloom. The fescue's flowers are relatively inconspicuous, and are just as blue as the leaves. The design incorporates an innovative water feature installed below the clear paving in the foreground.

Oriental gardens

Chinese and Japanese gardens are oases of green in which rocks and gravel, water, and plants are combined to create idealized pictures of nature. The effect relies on the juxtaposition of foliage shapes and textures, making bamboos and grasses perfect choices.

Pictures clockwise from below

Bamboo grove If space is not limited, bamboos such as *Yushania maculata* can be planted alone and allowed to form groves, through which paths can be cut.

Contemplative scene The main impact of this Japanese-inspired garden comes from foliage colour and shape, including the fine leaves of grasses and sedges, which are used primarily around the pond.

Oriental screen Traditional wooden Japanese-style screens provide the perfect backdrop to clumps of *Phyllostachys*. In Japan, the screens would be faced with opaque heavy rice paper; in colder and wetter climates, clouded Perspex is more appropriate.

Authentic composition A shady corner in this Japanese tea garden suits arrow bamboo (*Fargesia*), which displays its elegant, arching habit well, echoing the pendulous *Wisteria* flowers and leaves to left in the foreground.

Screens, hedges, and dividers

Grasses and bamboos are seldom seen as hedges or screens, but the larger species and cultivars, particularly evergreen bamboos, pampas grass, and the larger *Miscanthus* grasses, are ideally suited to this purpose.

Pictures clockwise from far left

East meets west Tall, evergreen, and growing at least as quickly as most coniferous hedges, bamboos can quickly create an impenetrable, elegant, and low-maintenance hedge. Here, green-stemmed *Phyllostachys* species planted closely together form an effective backdrop to a classical style bench and urn on a pedestal in an unusual east meets west juxtaposition of designs and influences.

Drawing the eye Positioned carefully, even a single bamboo can effectively mask an unwanted view or feature, such as a shed or dustbin, or act as the terminus to a more traditional conifer hedge, and become a focal point in its own right.

Large grass screens The tallest ornamental grasses can be used to form very effective hedges and screens if planted closely enough together so that the clumps merge. The largest *Miscanthus sinensis* cultivars, such as 'Grosse Fontäne' and 'Graziella', can top 1.8m (6ft), while *Miscanthus sinensis* var. *condensatus* 'Cosmo Revert' can reach 3m (10ft). All flower well over a long period, and the movement of their leaves and plume-like flowerheads in the wind adds an extra dimension of movement and sound. *Miscanthus* are deciduous grasses, but the bleached dead stems and flowerheads are attractive even in winter. For an evergreen grass screen, pampas grasses (*Cortaderia selloana*) are ideal, again reaching up to 3m (10ft) high.

Meadow dividers For a low screen that divides areas but can also be looked over as well as through, "exotic meadows" of small grasses mixed with perennials are the perfect solution. Feather reed grass (*Calamagrostis*), hair grass (*Deschampsia*), and switch grass (*Panicum*) all grow to around 90–120cm (3–4ft), and look good planted alongside understated perennials, such as *Arisaema*, *Aquilegia*, and *Astrantia*.

Container designs

Most grasses and grass-like plants thrive in containers and raised beds, bringing colour and movement to patios and decking. Many bamboos also look good in pots, and confining vigorous species in this way limits their spread.

Pictures clockwise from right

Harmonious patio Hard landscaping like paving and steps can look rather stark in a garden without the softening effect of plants. In this sophisticated scheme, gaps have been left in the lower course of paving to house grasses and sedges. The limited amount of soil to which these plants have access makes the gaps behave like pots sunk into the ground, which helps tie the containers on the step into the arrangement. The repeated use of gold-leaved sedge in a pot and against the lower step helps integrate the scheme, as do the blue-grey sets and grey stones surrounding the blue-leaved *Festuca glauca* grasses. Like the grasses and sedges, the clipped box is evergreen, providing year-round interest, while the pot of flowering violas introduces a seasonal summery note.

Uniform plantings Repetition of the same plant in the same container introduces formality into a design, particularly when both plants and containers are large. Here, three contemporary cubes house specimen black bamboos (*Phyllostachys nigra*), whose stems turn jet black with age. They act as a hedge or screen, marking the edge of the decking and giving some privacy from the overlooking windows. Bamboos in pots must not be allowed to dry out, so it is a good idea to install an automated irrigation system to keep them constantly moist.

Stainless steel beds A series of modern stainless steel raised beds in the corner of a decked area gives the opportunity for mixing and matching foliage colours and shapes. A bamboo in the background is underplanted with a purple-leaved New Zealand flax (*Phormium*) and tufty *Festuca glauca*. The blue-leaved hosta in the lower bed echoes the colour of the grass but has a contrasting leaf shape, while the slightly larger yellow hosta behind matches the yellow-variegated sedge in front and also provides an interesting mix of foliage forms.

Ground-cover plantings

Many grasses, grass-like plants, and a few of the smaller bamboos make excellent ground-cover plants. Space several plants at regular intervals and they will soon knit together into an ornamental, weed-suppressing block.

Pictures clockwise from left

Shaded elegance One of the most
beautiful of all grasses, *Hakonechloa macra*
is available in several variegated forms and
will grow happily in shade, although the
variegation is brighter in sun. Here, it forms
a carpet beneath a cherry tree, its arching
stems rippling in the slightest breeze.

Understated show This simple but
beautiful scheme includes clipped box,
yellow-flowered spurge (*Euphorbia*) and
tufted hair grass (*Deschampsia cespitosa*).
Deschampsia is an excellent clump-forming
grass for planting in shade or semi-shade.
It produces airy flowerheads relatively early
on in the year.

Damp shade duo Few grasses relish
damp, heavy soils in shade but, fortunately,
many sedges thrive in such conditions. Here,
gold-edged hostas are interplanted with
a low, sprawling mat of golden sedge
(*Carex aurea*).

Black and gold mix This simple but
effective ground-cover combination sets
gold variegated *Hakonechloa macra* with
the black grass *Ophiopogon planiscapus*
'Nigrescens'. One of few plants with truly
black foliage, this *Ophiopogon* is not a
grass at all but a member of the bindweed
(*Convolvulus*) family.

Bog and water gardens

Few grasses relish permanently wet feet, unlike sedges and other grass-like plants, which thrive in such conditions. Many of these are, in fact, true water plants that grow with their roots submerged but their leaves above water.

Pictures clockwise from top left

Container water garden Creating a water garden in a large container, such as a half-barrel, brings a water feature within the scope of even the smallest garden. Choose carefully, since many water plants are too vigorous for such a limited space, and confine them within plastic planting baskets. Here, upright *Typha minima*, the smallest of the reed maces, complements the floating leaves of a dwarf water lily (*Nymphaea* cultivar).

Woodland stream One of the few shade- and moisture-tolerant true grasses, hairgrass (*Deschampsia cespitosa*) is a native of woodland edges and clearings, forming attractive, sprawling green tussocks and flowering relatively early in the year. The wider and more upright leaves of Siberian irises in the shade of a nearby tree complement it well in this naturalistic streamside planting.

Luxuriant planting Around the edges of this small pond an abundance of foliage – grasses, sedges, and herbaceous plants – is used to great ornamental effect. The large grass on the right, *Cortaderia fulvida*, prefers a moist site. Although perennial, it performs like many annual grasses and begins to brown as flowering progresses, but the dead leaves are easily removed. In the foreground, the foliage of Bowles' golden sedge (*Carex elata* 'Aurea') and the steely-hued *Festuca glauca* form a colourful ensemble with the wide-leaved variegated hostas and upright yellow spikes of *Ligularia*.

Minimalist stripes Around the edge of an imposing concrete wall, which hides tanks designed to capture water to fill the pond, the strongly ascending stems of a horsetail, *Equisetum ramosissimum* var. *japonicum*, echo the stripes in the concrete. The horizontal bands formed by the horsetail nodes and the more loosely structured reeds (*Phragmites*) on the right make a striking contrast, and help to create a simple, elegant, arresting contemporary design.

Getting started

To understand how to grow and care for your plants, it's useful to know how they live in their natural habitats. In this chapter, find out more about bamboos, grasses, and other plants, so that you can provide them with the conditions they enjoy. Also learn how to evaluate your site and soil, and follow the tips on choosing a planting style to suit your garden. Finally, get your plants off to a good start with some simple planting and sowing techniques.

What is a grass?

Grasses form one of the largest and most widely distributed families of flowering plants. All true grasses are members of the *Poaceae*, but not every plant with strap-shaped leaves is a grass.

Natural ground cover Grasses form the dominant vegetation across great swathes of the Earth, including the African savannah, North American prairies, Russian steppes, and South American pampas. They can survive on little water, and grow from the stem and leaf bases, so they are able to withstand damage, including grazing.

Food sources Almost all the world's most important crop plants are grasses, including rice, wheat, barley, oats, and rye. The grass family could be seen as the foundation of human society – it formed the basis of settled farming, including the establishment of pastures for raising animals.

Pollination Grass flowers are very different from the bright, nectar-rich, often scented blooms of most plants, because they are wind- not insect-pollinated. The male flower parts produce huge quantities of tiny, light pollen grains that are blown onto the female parts by chance.

Varied form and colour There are relatively few grasses with real ornamental qualities, given the size of the family. However, those that are decorative offer a wealth of colours, shapes, and sizes. Not all are green: grasses may be blue, red, bronze, or silver. There are annual species, perennials, deciduous grasses with good autumn colour, and many evergreens. Some are clump-forming, others spread sideways, forming ground cover.

Ethiopian fountain grass (*Pennisetum villosum*) produces beautiful white "foxtail" flowers. It forms spreading tussocks but is not that hardy.

Miscanthus sinensis 'Silberfeder' is one of the largest of a group of late-flowering ornamental grasses, capable of reaching 1.5m (5ft).

Imperata cylindrica 'Rubra' is a striking-looking grass, with red colouring that creeps down the leaves as the season progresses.

Elymus magellanicus, a blue wheat grass from South America, forms lax, intensely silver-blue clumps. Few plants can match it for colouring.

What is a bamboo?

Bamboos are simply grasses with the ability to make wood, allowing them to grow much larger than their herbaceous, non-woody family relatives. Most species are tropical or sub-tropical.

Would-be giants Some bamboos are forest giants, tens of metres tall, that compete with trees in tropical rainforests. Rather than expanding with age like a tree, bamboos instead throw up a large number of "trunks" from a root system that expands sideways. The spread of many species is theoretically infinite.

Root systems Bamboos have two basic growth forms, controlled by the root system. Clump-forming species expand sideways relatively slowly; their roots branch and produce new culms (stems) at short intervals. Running species send up new shoots from faster-spreading roots at wider intervals, and can be invasive. Most temperate species behave like clump-formers.

Clumping bamboos include:

- *Chusquea*
- *Fargesia*
- *Himalayacalamus*
- *Phyllostachys* (in behaviour)
- *Semiarundinaria* (in behaviour)
- *Thamnocalamus*

Running bamboos include:

- *Bashania*
- x *Hibanobambusa*
- *Indocalamus*
- *Pleioblastus*
- *Pseudosasa*
- *Sasa*
- *Sasaella*
- *Yushania* (in behaviour)

Clumping root system   Running root system

Bamboo shapes and forms Bamboos are foliage plants *par excellence* and have a unique presence in the garden. Most of them are nowhere near as invasive as is popularly believed, being no more difficult to control than trees or shrubs that require annual pruning. They range in size from true dwarf bamboos, only tens of centimetres high, to towering plants rivalling trees for stature. The diversity in height, leaf size, culm (stem) colour, and habit is always expanding as new species and cultivars are introduced into cultivation.

Fargesia robusta is a relatively large-leaved, exposure-tolerant species that makes a good hedge, reaching 4m (12ft). The canes have fawn sheaths.

Phyllostachys nigra is unique, and deservedly popular, for the contrast between its green leaves and the shiny culms, which turn black as they age.

Sasa palmata f. *nebulosa* is handsome, but it runs to excess and should be confined to a container or kept out of smaller gardens altogether.

Pleioblastus variegatus also runs to excess (like all *Pleioblastus*), but is one of the best variegated bamboos, and much smaller than *Sasa*, so easier to control.

Plants that resemble grasses

A wide range of plants with long, narrow, strap-like leaves are often called grasses, even though they are unrelated to the grass family. Here are some of the best of the grass-like plants.

Moisture-loving grass-like plants Many grass-like species, such as reedmace (*Typha*), thrive in moist, or downright wet, soils and in full or partial shade – conditions that the fine, fibrous root systems characteristic of true grasses do not tolerate.

Different families and forms

It is difficult to generalize about plants resembling grasses, since most are not remotely related to the grass family, or to each other (as a close examination of their flowers will reveal). With some, such as many of the sedges, and the *Carex* genus in particular, their similarity to grasses is so close it is difficult to tell them apart, whereas with mondo grass, the flowers are so obviously designed to attract insects that, in bloom, it is clear they are not grasses at all.

Sedges The sedge family is quite large and diverse, and in evolutionary terms, much older than the grass family. Most of its members are evergreen and found in wet ground.

Woodrushes *Luzula* are members of the rush family adapted to life in the woods. They are very drought- and shade-tolerant, and one of the few grass-like plants to thrive under trees.

Rushes True rushes of the genus *Juncus* are real moisture-lovers, with upright stems and leaves so reduced as to seem absent. Cultivars of several species have curly leaves.

Reedmace *Typha* species form their own family and are often, incorrectly, called "bulrushes". Handsome aquatic plants, they are vigorous growers and can prove invasive.

Horsetails *Equisetum* are ancient, primitive plants, which reproduce like ferns, by spores and not seeds. They are best grown in pots to curb their wandering tendencies.

Mondo grass Most often seen in its black-leaved form, *Ophiopogon planiscapus* 'Nigrescens', mondo grass, produces small, lily-like flowers followed by black berries.

Evaluating your site

One of the secrets of success in gardening is putting the right plant in the right place, but growing conditions can vary markedly even within the smallest plot, so check yours carefully.

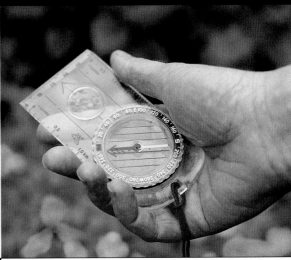

Which way? The direction a garden faces is known as its aspect, and is important because it governs the amount of direct sunlight the garden receives. South-facing gardens get the most sun, north-facing ones the least. Those facing east or west get full sun for part of the day. It is easy to work out your garden's aspect with a compass.

Weak morning sun illuminates most of this garden.

Sun, shade, and wind A garden's immediate surroundings also affect the amount of light it receives. Buildings, hedges, and large trees all get in the way of direct sunlight, casting shadows and creating shade. Even south-facing gardens surrounded by buildings or trees can be in shade for most of the day. Sunlight has a major bearing on which plants can be grown successfully in a garden, but another important factor is exposure – in an open, sunny garden without walls or hedges, there is nothing to break the prevailing wind, which can damage plants and suck moisture out of leaves. Some plants can cope with this exposure, others prefer a more sheltered site where the effects of the wind are tempered by buildings, fences, or trees.

At midday, the sun is closest to overhead and there is little shade.   By evening, much of the garden is shaded by the hedge at the rear.

Assessing a garden's microclimates When deciding where to place a plant in your garden, look at the growing conditions, and the needs and preferences of the plant itself. Grasses tend to prefer an open, sunny spot, which is also likely to be drier, whereas sedges, woodrushes, and other grass-like plants prefer shadier conditions. Most bamboos will take sun or part-shade.

Assessing your soil

The type of soil in your plot determines which plants will grow happily there. Put a plant in a soil similar to that in which it grows in the wild, and it should do well, but give it the wrong soil, and it will fail to thrive and may even die.

Different soil types

Before you plant anything in your garden, take the time to assess your soil type – it is crucial when selecting appropriate plants. Unsuitable or poor soils can be improved (*see opposite*), but you have to know your soil type before choosing a method. Soil type can vary markedly across the same garden, too, so check the soil in several different areas.

Clay soil is made up of tiny particles, holds water well but can waterlog, and tends to be high in plant nutrients.

Sandy soil has much larger particles, drains freely and holds much less water, and is often poor in nutrients.

Silty soil has larger particles than clay but can still waterlog, and is reasonably fertile.

Loamy soil has a balance of particle sizes and organic matter, holds some water well and is fertile – ideal!

Testing for pH A soil's pH indicates its basic chemistry, whether it is acid, alkaline, or neutral. Expressed on a scale from 1 to 14, 1 is extremely acid, 14 very alkaline. The midpoint, 7, is neutral. A soil's pH is important because most plants have a preferred range. Some will grow well only in acid conditions, while others need alkaline soils, like chalk or limestone. Most grasses and bamboos are fine in soils around neutral, but some moisture-lovers prefer acid soils, and some grasses really prefer alkaline conditions. Use a soil test kit (*below*) to check the pH of your soil, and sample different areas – borders by walls are often more alkaline, for example.

Soil texture Soil is made up of particles whose size controls the amount of water, organic matter, and plant nutrients it can hold. To assess soil texture, dig a small sample of soil from the garden and rub it between your fingers. If it feels gritty and will not form a ball, but leaves your hands fairly clean, it contains a lot of sand. This soil is likely to be free-draining and on the dry side, and many grasses prefer this. If a soil feels smooth, slippery and sticky, and can be moulded into a ball, it contains a lot of clay or silt, will hold moisture well and is likely to be fertile, but will waterlog fairly easily (when it will hold little or no air). Sedges, rushes, and reedmaces enjoy such conditions.

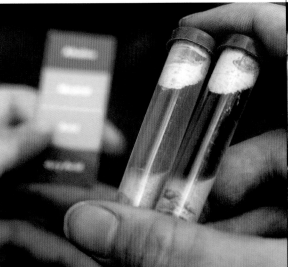

The left-hand tube contains near-neutral soil, the other is alkaline.

Sandy soil is crumbly and dry.   Clay soils will roll into a ball.

Adding organic matter to improve a sandy soil

Completely changing a soil's texture and pH is impossible but most soils can be improved and their worst features reduced. This may be necessary only in parts of the garden. The best way to improve the moisture-holding capacity and nutrient content of a sandy soil is to add plenty of bulky organic matter, such as garden compost, composted bark, or well-rotted manure. As well as holding moisture, the organic matter breaks down in the soil, releasing nutrients and helping to stick the soil particles together, improving its overall structure. Initially, dig the organic matter into the soil, but once an area has been planted, you can subsequently apply it every year as a surface mulch.

Lightening a heavy soil Heavy clay soils contain plenty of nutrients but often hold too much moisture, resulting in waterlogging in wet weather. Few plants relish permanently wet feet, and grasses in particular can rot in such conditions. To improve the drainage in a clay soil, dig in plenty of grit and/or sand. This helps to break up the clay, encouraging it to form a better "crumb" structure, creating spaces, or pores, through which water can drain more easily. Adding organic matter (using the same method as for a sandy soil, *above*) at the same time also helps to improve the physical structure of a clay soil, which further aids drainage.

Removing topsoil to impoverish a rich soil

Generally, gardening is about creating better, more fertile soil, but in a few instances, the opposite is required. When establishing an annual or perennial meadow, with a mixture of grasses and wild flowers, the aim is to create conditions that favour the flowering plants over the grasses. On rich soils, grasses grow so vigorously that they crowd out the wild flowers, which are better adapted to growing on nutrient-poor soils. Scraping off the top 10–15cm (4–6in) of topsoil and using it elsewhere in the garden, then sowing a meadow mix onto the exposed, lower-nutrient subsoil, is an extremely effective way of producing a flower-rich annual or perennial meadow.

Choosing a planting style

The type of garden you create is down to personal taste. However, some sites are better suited to particular styles than others: prairie plantings, for example, do best on open, sunny sites, while bog gardens need cool, damp conditions.

Meadow and prairie planting

Both meadows and prairies mix grasses with herbaceous plants, but while prairies are natural communities, meadows are a creation of traditional farming practices. Both styles look natural and establish quite quickly.

Pros and cons Both styles are low-maintenance, easy to establish, suitable for all garden sizes, and encourage wildlife. They are not suitable for shady or boggy ground. Annual meadows need resowing every spring and have little late-season interest; prairies peak in late summer.

Garden requirements Open, sunny sites with free-draining, relatively infertile soils suit prairies and meadows.

Suitable plants:

Prairies – large grasses
- *Calamagrostis* x *acutiflora* cultivars
- *Cortaderia selloana*
- *Miscanthus sinensis*
- *Stipa gigantea*

Meadows – smaller grasses
- *Agrostis* species
- *Briza maxima*, *B. minor*
- *Deschampsia cespitosa*
- *Festuca* species
- *Molinia caerulea*

Bog, water, or shade planting

Many sedges and other grass-like plants, such as rushes, reedmaces, and horsetails thrive in wet conditions – some can be grown in ponds. Few grasses or bamboos relish wet feet, but many bamboos cope with dry shade.

Pros and cons Lush effects can be created in bog gardens and by ponds, mixing sedges and grass-like plants with other herbaceous moisture lovers, but many grow quickly and spread, and need cutting back periodically.

Garden requirements Moisture-loving plants are happiest on heavy clay soils. Some plants will stand full sun, but most prefer at least partial shade.

Suitable plants:

Bog and water gardens:
- most sedges, especially *Carex* and *Uncinia*
- *Equisetum* (horsetails)
- *Juncus* (rushes)
- *Phragmites* (reeds)

Drier shade:
- many bamboos, especially *Fargesia*, *Indocalamus*, and *Pleioblastus* species
- *Luzula* (woodrushes)

Sowing a meadow from seed is a cheap way of creating a pretty garden quickly. Large areas can look stunning.

Permanently moist areas are often seen as problematic, but with the right plants they can be transformed into beautiful areas.

Contemporary, modern, or containers

Bamboos, grasses, and grass-like plants suit modern gardens extremely well, from sparse, minimalist designs to contemporary "outdoor rooms", where decking or a patio may replace the traditional lawn. Containers can be used in any style of garden, and many of the species featured in this book look great and perform well in pots.

Pros and cons The clean lines of modern materials like metal and glass complement the linear foliage and upright lines of bamboos and grasses beautifully. The plants also suit containers well, particularly when grouped. Container-grown plants need regular watering and, if permanently planted, regular dividing and repotting.

Garden requirements Courtyards, town, and roof gardens suit these styles best, but almost any space can be given a fresh edge with good landscaping. With containers, you are not constrained by soil or garden type; there are pots and plants to suit all situations.

Suitable plants:

- many bamboos, eg, *Phyllostachys* species
- most sedges, especially *Carex* and *Uncinia*
- *Calamagrostis* x *acutiflora* cultivars
- *Cortaderia selloana* 'Pumila'
- *Elymus magellanica*
- *Fargesia* species
- *Festuca glauca*
- *Pleioblastus* species

Adding to traditional borders

Gardeners can be shy of using grasses in traditional mixed borders, but a few well-chosen grasses or a single bamboo can add much to herbaceous plantings and beds that include shrubs. Their movement and shape provide attractive contrasts, and just as with prairie plantings, the combination is not seen as odd at all. There is also a range of foliage colours to choose from.

Pros and cons Fascinating contrasts in colour and shape can be achieved by introducing grasses and bamboos to traditional gardens, but position them with care – don't add so many that the original style is lost. Repeating the same species at intervals along a border can be particularly successful.

Garden requirements Virtually any size and style of garden, no matter how traditional, can be improved with the addition of a few grasses, and arguably with the inclusion of an exotic-looking bamboo or two.

Suitable plants:

- many sedges, eg, *Carex elata* 'Aurea' and *Carex oshimensis* 'Evergold'
- *Cortaderia selloana*
- *Deschampsia cespitosa*
- *Fargesia robusta*
- *Luzula sylvatica*
- *Phalaris arundinacea* var. *picta*
- *Pleioblastus viridistriatus*
- *Stipa gigantea*
- *Thamnocalamus* species

Simple plantings using a restricted colour palette and repeated clumps of the same species create a contemporary feel.

There is a wealth of decorative grass species suitable for all types of beds and borders, and all sizes of garden.

Planting out a container-grown grass

Most grasses are bought in containers, and are very easy to plant. The best planting times are spring, allowing the grasses a full season to establish, or autumn, when the soil is usually wet enough for root growth before winter.

1 Dig a hole in the ground large enough to take the full depth of the container. Place the potted plant in the hole to ensure that the top of the compost is level with the soil. Water the grass well in its pot before planting.

2 Remove the plant from its pot and gently tease out the roots around the edges of the root ball. This encourages them to grow out into the surrounding soil.

3 Place the plant centrally into the hole, and fill in with the excavated soil around the edges, firming gently with your fingers. Some gardeners like to mix compost into the excavated soil, but this is not absolutely necessary.

4 Using a watering can with a fine rose, gently water the plant. Watering ensures that soil particles are washed into close contact with the roots. Keep the plant moist while it becomes established.

Dividing grasses

Large grasses can often be divided at planting-out time, yielding extra plants for free. Wait until late summer to divide *Miscanthus* species and cultivars though; they react badly to being divided earlier.

Tip for success

Leave the crown of newly split grasses slightly proud of the soil surface to discourage the development of root rots.

1 Remove the plant from the pot and look at the natural lines of growth to see where the plant (here *Leymus arenarius*) is best sliced. Larger specimens may yield as many as three or four divisions.

2 Using a saw or large serrated knife, cut through the crown of the plant along the identified division line. The crown is usually not more than a few centimetres deep, and cutting will become easier once you have sliced through it.

3 Once the crown has been cut through, gently prise apart the rest of the root ball with your hands – this actually causes less damage to the fibrous grass root system than cutting it.

4 Plant out the divided pieces as you would container plants (*see pp.44–45*), spacing them at least 30cm (12in) apart and watering them in well. Three or more clumps create an instant effect.

Planting bamboo in a container

Bamboos can look extremely attractive growing in containers, adding an exotic oriental touch to patios or decking, and providing a bold contrast to other plants in pots. They are easy to look after as long as they are not allowed to dry out.

1 Choose a container that complements the bamboo (glazed oriental pots look particularly good) and is large enough to give the plant room to spread. Place plenty of crocks in the base of the pot to aid drainage.

2 Put some loam-based, John Innes No.3 compost into the bottom of the container and check that the surface of the root ball sits 3–4cm (1¼–1½in) below the pot rim (the bamboo shown here is *Pleioblastus variegatus*).

3 Take the bamboo out of its pot. Loosen the root ball (a mix of roots and thicker rhizomes) to encourage the roots to grow into the soil. Cut off any rhizomes that are too long to fit easily into the pot.

4 Place the plant in the centre and put compost around the edges, firming gently with the fingers. Ensure the root ball is covered with 2–3cm (¾–1¼in) of compost, because bamboos tend to "heave" upwards in pots over time.

Planting a bare-root bamboo

Bamboo nurseries grow most of their stock plants in the ground, and supply plants by mail order when they are dormant, with little or no soil. Plant out bare-root stock as shown here.

Tip for success

Firm between stems to make sure the plant is steady. If any stems move too much, keep them upright with short stakes.

1 Wrap the bare-root plant in plastic to keep it moist and don't remove until the last possible moment; exposed roots dry out and die very quickly. Dig a hole deep enough to accommodate the root ball.

2 Mix the excavated soil and the soil at the base of the hole with a few generous handfuls of garden or potting compost, or composted bark.

3 Unwrap the bamboo and sit it on a layer of the soil and compost mix, checking that the marks on the culms showing the previous soil level are at, or slightly below, the ground surface.

4 Keeping the plant upright with one hand, add more of the compost and soil mix, working it into every cranny. It is important not to leave any air holes, but take care not to damage the plant. Fill to ground level, firm, and water in.

Sowing annual grasses

Annual grasses are often sown where they are to grow and flower, but the best way of growing attractive annual species like this *Briza maxima* is to sow seed in modules each spring.

Tip for success

Keep any newly-sown trays in semi-shade while the seedlings are germinating, then move them into full sun.

1 Fill a seed or module tray with a low-nutrient seed compost, such as John Innes No.1, to within half a centimetre of the lip, and tap the tray gently to settle the compost. Top up if necessary.

2 Carefully sow three or four seeds per module onto the surface of the compost. This quantity of seeds should produce nice chunky little clumps of grass.

3 Sprinkle a thin layer of vermiculite on top of each module. Vermiculite keeps the compost surface moist and protects seeds from drying breezes, while still allowing light through to encourage germination.

4 Water the tray well with a fine rose, and keep the compost moist. Depending on the grass species, germination may take a day or up to three weeks. Pot on or plant out when the seedlings are about 15cm (6in) high.

Sowing a meadow

With a good-quality seed mix and not-too-rich, well-prepared soil, both annual and perennial meadows are remarkably easy to establish. Very low maintenance, they can be created on any scale.

Tip for success

Protect newly sown areas from birds by stringing CDs on a line; they flash in the light, making effective high-tech "scarecrows".

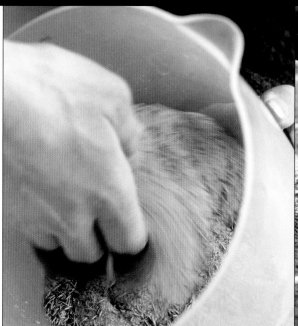

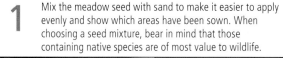

1 Mix the meadow seed with sand to make it easier to apply evenly and show which areas have been sown. When choosing a seed mixture, bear in mind that those containing native species are of most value to wildlife.

2 Prepare the seedbed carefully; remove perennial weeds and rake the soil flat to a fine crumbly texture. Using canes, mark out a grid to help you to distribute the seeds evenly, and sow at the supplier's recommended rate.

3 Once the area is sown, press it down gently with the back of a rake to ensure the seed is in contact with the soil. Be careful not to bury the new seed because light is important for the germination of most meadow species.

4 Grasses and wild flowers should be growing vigorously in two to three weeks. Some annual mixes may be in full flower in only six weeks. Perennial mixes are slower to establish; some may take two years to reach their peak.

Planting an invasive aquatic plant

Marginal plants such as *Phragmites, Typha,* and *Equisetum* can be invasive, so in small ponds and bog gardens it is best to contain them in special baskets, which can be placed in the pond, or in pots that can be sunk into the soil.

Tip for success

Choose a plastic or netting mesh basket that allows plenty of room for growth, but is not too large for the planting site.

1 Half-fill the basket with ordinary garden soil or a specialist aquatic compost. Remove the plant from its original pot and check that there is enough room for the plant and a surface mulch of gravel, approx 2–3cm (¾–1¼in) deep.

2 Place the plant centrally in the basket and fill around the root ball with more soil or compost, firming gently and making sure that the plant is upright. The plant shown here is a horsetail, *Equisetum hyemale* var. *affinis*.

3 Top up the basket with a mulch of gravel or small stones, which will help retain the compost. This is particularly important if your pond contains fish because they are inclined to nose around in compost.

4 Lower the basket carefully into the pond, ensuring it sits below the water level. The existing vegetation in this pond needs thinning out and illustrates the vigour with which many water plants colonize open water.

Planting recipes

The beautiful planting combinations in this chapter are easy to achieve, and the cultivation tips show you how. You can also easily adapt the recipes to suit your own garden or planting style. The symbols below are used in the recipes to indicate the conditions the plants prefer.

Key to plant symbols

♈	Plants given the RHS Award of Garden Merit

Soil preference

◌	Well-drained soil
◑	Moist soil
●	Wet soil

Preference for sun or shade

☼	Full sun
◐	Partial or dappled shade
◉	Full shade

Hardiness ratings

✳✳✳	Fully hardy plants
✳✳	Plants that survive outside in mild regions or sheltered sites
✳	Plants that need protection from frost over winter
❀	Tender plants that do not tolerate any degree of frost

All-grass border

Borders composed entirely of grasses are very low maintenance and suited to both contemporary and more traditional gardens. They rely for their effect on a combination of contrasts: height, plant habit (upright or arching), leaf size and colour, and flower shape and colour. Chosen well, grass borders can be very colourful with a long season of interest, and their movement brings an extra dimension to the garden. Here, the upright forms of *Stipa calamagrostis* and *Miscanthus sinensis* 'Zebrinus' create the backbone, with the arching flowerheads of *M. sinensis* 'Morning Light' in between. Clumps of smaller grasses make up the foreground.

Border basics

Size 3x1.5m (10x5ft)

Suits Most gardens, particularly contemporary

Soil Free-draining

Site Full sun to limited shade

Shopping list

- 2 x *Stipa calamagrostis*
- 1 x *Miscanthus sinensis* 'Zebrinus'
- 1 x *Miscanthus sinensis* 'Morning Light'
- 2 x *Pennisetum alopecuroides*
- 3 x *Hakonechloa macra* 'Aureola'
- 1 x *Elymus magellanicus*

Planting and aftercare

Position the taller species towards the back of the border, allowing at least 45cm (18in) between them since they all form substantial clumps. Place the smaller grasses towards the front, grouping the three slow-growing *Hakonechloa* about 20cm (8in) apart. Add a gravel mulch to reduce weeds and help conserve water. Water for the first year while the planting establishes. Although deciduous, these grasses can be left to provide winter structure.

Stipa calamagrostis
❄❄❄ ◊ ◊ ☼ ☼

Miscanthus sinensis 'Zebrinus'
❄❄❄ ◊ ◊ ☼ ♀

Miscanthus sinensis 'Morning Light'
❄❄❄ ◊ ◊ ☼ ♀

Pennisetum alopecuroides
❄❄❄ ◊ ◊ ☼

Hakonechloa macra 'Aureola'
❄❄❄ ◊ ☼ ☼ ♀

Elymus magellanicus
❄❄/❄❄❄ ◊ ◊ ☼

Grasses in a contemporary border

Some grasses look beautiful in mixed borders alongside herbaceous plants. One of the best large grasses for such a scheme is Spanish oat grass (*Stipa gigantea*), with its delicate, almost transparent, long-lasting flowerheads up to 1.8m (6ft) tall that hover above lush clumps of foliage around 90cm (3ft) high. In this simple but very effective scheme, its flowers form an airy screen, linking the canopy of birch trees (*Betula*) to the lower planting of white geraniums and red granny's bonnets (*Aquilegia*). The feathery foliage of fennel (*Foeniculum*) echoes the grass flowers.

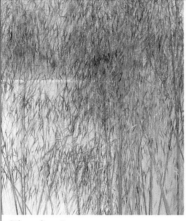

Stipa gigantea
❁❁❁ ◊ ☼ ♀

Betula utilis var. *jacquemontii*
❁❁❁ ◊ ◕ ☼ ◑

Border basics

Size 3x1.5m (10x5ft)

Suits Modern or cottage-garden border

Soil Most, except excessively wet or dry

Site Full sun to partial shade

Shopping list

- 4 x *Stipa gigantea*
- 2 x *Betula utilis* var. *jacquemontii*
- 8 x *Aquilegia vulgaris* var. *stellata* 'Ruby Port'
- 4 x *Geranium phaeum* 'Album'
- 6 x *Foeniculum vulgare* or *F. vulgare* 'Purpureum'

Aquilegia vulgaris var. *stellata* 'Ruby Port' ❁❁❁ ◊ ◕ ☼ ◑

Geranium phaeum 'Album'
❁❁❁ ◊ ◕ ☼ ◑

Planting and aftercare

Plant the birch trees first, at regular intervals, towards the back of the border, with a line of *Stipa* snaking between them. Then add the geraniums towards the front, with the fennel and aquilegias dotted randomly around the remaining unplanted area for a natural feel, as if they had self-seeded.

By midsummer, the aquilegias and geraniums will cease flowering and the fennel will be as tall as the *Stipa*. For a moodier look, replace the green fennel with its purple-leaved cultivar *F. vulgare* 'Purpureum'. This planting works equally well in full sun or partial shade.

Alternative plant idea

Foeniculum vulgare
❁❁❁ ◊ ☼ ◑

Foeniculum vulgare 'Purpureum'
❁❁❁ ◊ ☼ ◑

Japanese influences

A Japanese-style garden represents an idealized form of nature, with a restful atmosphere that changes relatively little through the seasons. The emphasis is on foliage and the contrasts of different leaf colours, sizes, and textures. Rocks and gravel are often prominent, with ornamentation, such as a lantern, understated but very carefully placed. Here, the lantern in the background, surrounded by foliage, terminates the view and is the lynchpin of the whole composition without dominating it. Other suitable plants for a Japanese-style garden include sedges, ferns, hostas, and evergreen azaleas.

Border basics

Size 3x6m (10x20ft)

Suits Oriental-style gardens

Soil Moist and fertile

Site Partial shade

Shopping list

- 1 x *Pleioblastus variegatus*
- 1 x *Phyllostachys bambusoides* 'Castillonii'
- 2 x *Fargesia nitida*
- 1 x *Petasites japonicus*

Planting and aftercare

The tall *Phyllostachys* is planted close to the path to overhang it, while two finer-leaved *Fargesia* are beyond and to either side of the lantern, framing it with their hanging foliage. The two other key plants – large-leaved *Petasites* and short *Pleioblastus variegatus* – are staggered either side of the path. Together, these principal plants lead the eye from left, to right, left again, and right to rest finally on the lantern.

Keep plants moist while establishing, and groom them regularly to remove dead leaves. The *Petasites* and bamboos spread over time and need pruning to keep them in check.

Pleioblastus variegatus
❄❄❄ ◊ ◊ ☼ ☼ ⚲

Phyllostachys bambusoides 'Castillonii' ❄❄❄ ◊ ◊ ☼ ☼

Fargesia nitida
❄❄❄ ◊ ◊ ☼ ☼

Petasites japonicus
❄❄❄ ◊ ◊ ☼ ☼

Bamboo grove

Where space is available to let bamboos grow tall and spread sideways, you can create an oriental-looking grove or forest. Planting several different bamboos together allows subtle contrasts between leaf size, plant shape, and stem (culm) colour to come through. The *Phyllostachys violascens*, the largest species here, and *Thamnocalamus* are grown primarily for their stem colour, while *Fargesia robusta* is a good, densely clumping, all-round bamboo.

Border basics

Size 5x5m (15x15ft) – but will spread beyond this if not managed
Suits Medium-sized to large, oriental-style or woodland gardens
Soil Free-draining to moist

Shopping list

- 1 x *Thamnocalamus spathiflorus*
- 3 x *Pleioblastus viridistriatus*
- 1 x *Phyllostachys violascens*
- 1 x *Fargesia robusta*

Planting and aftercare

For immediate impact, buy large specimens of these bamboos. They can still take a year or two to begin growing away strongly – this forest effect is not achieved overnight.

Create a pathway with bark chips or gravel, and plant the three large bamboos at least 2m (6ft) apart, to the back and on either side of the path, with the smaller *Pleioblastus* spaced evenly in front of them. The planting will look sparse to begin with, but to help disguise this, use temporary "filler" plants, such as grasses or annual bedding, and remove them as the bamboos spread. Once established, the only maintenance a bamboo grove should require is periodic thinning of old culms.

Thamnocalamus spathiflorus
✽✽✽ ◌ ◗ ☼

Pleioblastus viridistriatus
✽✽✽ ◌ ◗ ☼ ☀ ♀

Phyllostachys violascens
✽✽✽ ◌ ◗ ☼ ☀

Fargesia robusta
✽✽✽ ◌ ◗ ☼ ☀

Large-scale prairie planting

This popular style of low-maintenance gardening mimics the blend of large grasses and herbaceous plants of North American prairies and Russian steppes. It is particularly effective with large plants that are given plenty of space. The backbone of this border is provided by grasses, including the variegated giant Provencal reed (*Arundo donax* var. *versicolor*) and *Stipa gigantea*, but *Miscanthus* and *Calamagrostis* are just as suitable. Flower colour comes from *Achillea*, *Sedum spectabile*, and black-eyed Susan (*Rudbeckia*). Like the grasses, these perennials flower late in the season.

Border basics

Size 5x3m (15x10ft)

Suits Informal contemporary and wildlife gardens

Soil Free-draining

Site Full sun, wherever possible

Shopping list

- 1 x *Arundo donax* var. *versicolor*
- 3 x *Stipa gigantea*
- 3 x *Achillea filipendulina* 'Gold Plate'
- 7 x *Rudbeckia fulgida* var. *sullivantii* 'Goldsturm'
- 2 x *Sedum spectabile*
- 5 x *Calamagrostis brachytricha*

Planting and aftercare

Set out the plants in their pots and try to make their positioning look as random as possible, ideally forming small groups rather than dotting individuals about. Once established, prairie plantings are extremely drought-tolerant. Although most of the plants are deciduous, their flowerheads and stems stand up well to winter weather and look attractive rimed with frost, while their seedheads are appreciated by the local wildlife. Cut plants back to ground level in early spring, if not before, to give new growth room to come through.

Arundo donax var. *versicolor*
❄❄ ◐ ◑ ☼ ☀

Stipa gigantea
❄❄❄ ◐ ☼ ♈

Achillea filipendulina 'Gold Plate'
❄❄❄ ◐ ◑ ☼ ☀ ♈

Rudbeckia fulgida var. *sullivantii* 'Goldsturm' ❄❄❄ ◐ ◑ ☼ ☀ ♈

Sedum spectabile
❄❄❄ ◐ ◑ ☼ ☀ ♈

Calamagrostis brachytricha
❄❄❄ ◐ ◑ ☼ ☀

Small-scale prairie planting

The grass backbone of this beautiful prairie planting is supplied by *Stipa tennuissima* and *Anemanthele lessoniana*, both more compact, fine-leaved, and airy-flowered choices than *Calamagrostis* or *Miscanthus* cultivars. They are teamed with "classic" prairie perennials in a limited colour palette: orange-red *Helenium*, yellow yarrow (*Achillea*), and pale orange coneflower (*Echinacea paradoxa*) – *E. purpurea* would work equally well but would introduce another colour. The secret of success is to plant randomly so the grasses and flowers intermingle.

Border basics

Size 1x1m (3x3ft) or a large container

Suits Most styles of garden

Soil Free-draining

Site Full sun, ideally

Shopping list

- 2 x orange-red *Helenium*
- 2 x *Stipa tennuissima*
- 1 x *Anemanthele lessoniana*
- 2 x *Echinacea paradoxa* or *E. purpurea*
- 1 x *Achillea* 'Taygetea'

Planting and aftercare

Set out the grasses first, in their pots, then add the perennials, trying to make the planting appear as random and natural as possible, and grouping the plants quite closely together. Place the taller plants towards the back, with the shorter ones near the front.

These plants do well in dry conditions once established, but if in a container, they will probably need watering at least once a week. Apart from deadheading the perennials to prolong their flowering display, maintenance is minimal: cut back plants hard as they die off in autumn, or leave until spring so that you can enjoy their skeletal structure over winter.

Helenium
❋❋❋ ◊ ◗ ☼ ☀

Stipa tennuissima
❋❋❋ ◊ ◗ ☼ ☀

Anemanthele lessoniana
❋❋/❋❋❋ ◊ ◗ ☼ ☀

Echinacea paradoxa
❋❋❋ ◊ ◗ ☼ ☀

Achillea 'Taygetea'
❋❋❋ ◊ ◗ ☼ ☀

Alternative plant idea

Echinacea purpurea
❋❋❋ ◊ ◗ ☼ ☀

Meadow planting

Mixing smaller grasses with wildflowers, whether annual or perennial, is an increasingly popular form of low-maintenance gardening. Ornamental meadows essentially mimic traditional hay meadows, most of which have now been lost through the intensification of farming methods. Meadow plants are not only beautiful but they are also great for attracting wildlife into the garden. The planting style is extremely flexible, too, and can be used on any scale, from a large container to a whole field.

Avena sativa (wild oat)
❀❀❀ ◊ ☼

Centaurea cyanus (cornflower)
❀❀❀ ◊ ☼

Garden basics

Size Any size

Suits Informal, contemporary, and wildlife gardens

Soil Free-draining and not too rich

Site Full sun

Shopping list (seeds)

- *Avena sativa* or other annual grasses
- *Centaurea cyanus* or *Agrostemma githago*
- *Papaver rhoeas*
- *Leucanthemum vulgare*
- *Anthemis arvensis*

Planting and aftercare

The easiest and most effective way of establishing both annual and perennial meadows is to sow seed onto bare, prepared soil in spring. You can buy commercial seed mixes from specialist suppliers, or blend your own from flowers and grasses that appeal. Sowing the seed directly gives the random mix of species found in hay meadows.

Sow the seed evenly and thinly. Take out any obvious weeds, such as dock, nettle, and thistles, as they come through. Keep the germinating seedlings moist initially. After that, there is little to do other than cutting plants down to ground level in autumn and composting the dead material. Annual grasses must be re-sown every spring.

Papaver rhoeas (common poppy)
❀❀❀ ◊ ☼

Leucanthemum vulgare (ox-eye daisy)
❀❀❀ ◊ ☼

Alternative plant idea

Anthemis arvensis (corn chamomile)
❀❀❀ ◊ ☼

Agrostemma githago (corn cockle)
❀❀❀ ◊ ☼

Summer gravel garden

Grasses, particularly large ones, can look superb planted in gravel, which shows off their structure and habit extremely well. For this scheme, use a mixture of grasses with contrasting sizes, shapes, and flowers. Here, *Stipa gigantea*, *S. tenuissima*, and *Calamagrostis* are in full flower, with the *Miscanthus* planted in the rear yet to come into flower.

Border basics

Size 4x3m (12x10ft)

Suits Informal and contemporary gardens

Soil Free-draining

Site Ideally, full sun

Shopping list

- 1 x *Calamagrostis* x *acutiflora* 'Karl Foerster'
- 3 x *Miscanthus sinensis*
- 1 x *Stipa gigantea*
- 4 x *Stipa tenuissima*

Planting and aftercare

Before planting, lay a permeable landscape fabric over the area to be planted to help stop weeds appearing. Cut slits in a cross shape in the fabric where you want the grasses, then plant them through the holes. If you want the gravel to remain visible around the larger grasses, plant them at least 90cm (3ft) apart, but if you want a large block of the same cultivar, as here with the *Miscanthus*, position the plants more closely together. Spread the gravel at least 5–7cm (2–3in) deep over the fabric. *Stipa tenuissima* makes an excellent foreground plant, and although *Stipa gigantea* is a large grass, it is so airy that it also works well towards the front.

Gravel gardens are inherently low maintenance so, once established, this planting will only need cutting once a year in early spring.

Calamagrostis x *acutiflora* 'Karl Foerster' ✼✼✼ ◊ ☼

Miscanthus sinensis ✼✼✼ ◊ ◗ ☼ ◑

Stipa gigantea ✼✼✼ ◊ ☼

Stipa tenuissima ✼✼✼ ◊ ◗ ☼ ◑

Foliage forms for small spaces

This delightful little grouping is equally suited to a gap in decking, a large container, or small raised bed, and relies on the juxtaposition of different leaf shapes and colours for its effect.

Blue fescue (*Festuca glauca*) and New Zealand hook sedge (*Uncinia rubra*) contrast with the small-leaved variegated hosta and feathery flowers of *Astilbe*. The hook sedge is echoed in the wider leaves of a red *Phormium* 'Bronze Baby' (which will grow too large for the planting in a couple of years and need removing). Finally, the crinkly green leaves of the fern contrast well with those of all the other plants.

Bed basics

Size 90x45cm (3x2½ft)

Suits Contemporary and informal gardens, or containers

Soil Moist but free-draining

Site Sun or shade

Shopping list

- 1 x pink-flowered *Astilbe*
- 1 x *Phormium* 'Bronze Baby'
- 1 x *Uncinia rubra*
- 1 x *Festuca glauca*
- 1 x *Woodsia polystichoides*
- 3 x *Hosta* 'Patriot'

Planting and aftercare

This scheme is one for small spaces, so make sure that you position the plants closely together. Put the central plants in first, then work outwards, placing the hostas and fern so that they lean slightly out and spill over the edges of the decking. Finish off with a mulch of bark chippings. The hostas, fern, and hook sedge relish moist soil, while the fescue and *Phormium* prefer drier conditions, so water carefully, aiming for evenly moist, but never wet, soil. Trim back any plant as soon as it appears to be outgrowing its allotted space.

Astilbe
✿✿✿ ◐ ● ◌ ☼ ◑

Phormium 'Bronze Baby'
✿✿✿ ◌ ◐ ☼ ◑

Uncinia rubra
✿✿✿ ◐ ● ◑ ◑

Festuca glauca
✿✿✿ ◌ ◐ ☼ ◑

Woodsia polystichoides
✿✿✿ ◐ ◑ ♚

Hosta 'Patriot'
✿✿✿ ◐ ☼ ◑

Striking combinations

This planting shows off perfectly the architectural qualities of a large bamboo clump in a setting where space is not at a premium. The bamboo is complemented by the understated, low-growing plants beneath it.

Underplanting a substantial bamboo clump helps to keep weeds down and gives it a foil, but choose companions carefully because the ground at the base is usually dry. Avoid strong foliage or flower colours to leave the emphasis squarely on the bamboo itself, particularly if you prune the lowest branches to show off the culms (stems), as here. Some of the culms in this species are curiously crooked towards their base, adding extra interest.

Border basics

Size 3x1.5m (10x5ft)

Suits Informal, woodland, and contemporary gardens with space

Soil Free-draining to moist

Site Full sun to partial shade

Shopping list

- 1 x *Phyllostachys aureosulcata* or *Pseudosasa japonica*
- 1 x *Anemanthele lessoniana*
- 4 x *Erigeron karvinskianus*

Planting and aftercare

A similar arrangement could be made with any medium to tall bamboo species, in virtually any part of the garden where there is room. You could allow the clump to spread freely in an open situation at the edge of woodland, as here for example, or keep it in check with regular pruning. For immediate impact, purchase a fairly large bamboo. Plant it first, with the *Anemanthele* in front of it and the *Erigeron* (a daisy from Mexico that can be a little tender) spaced evenly around the foreground. Prune off the lowest branches of the bamboo, if desired.

Phyllostachys aureosulcata
❀❀❀ ◊ ◗ ☼ ☀

Anemanthele lessoniana
❀❀/❀❀❀❀ ◊ ◗ ☼ ☀

Erigeron karvinskianus
❀❀/❀❀❀❀ ◊ ◗ ☼ ☀ ♈

Alternative plant idea

Pseudosasa japonica
❀❀❀ ◊ ◗ ☼ ☀ ♈

Feather-topped screen

This simple yet effective planting creates a screen of grasses and sedges, with late-flowering perennials in front, and will flower from late summer right through the autumn.

Pampas grass (*Cortaderia selloana*) was overplanted in the 1970s and remains unfashionable, which is a pity because, *en masse*, it can make a wonderful hedge or screen. 'Sunningdale Silver' is a variegated selection with relatively compact flowerheads. Its grey-green leaves and white flowers contrast strongly with a substantial clump of red dahlias, which are teamed with the flat flowerheads of *Sedum* and *Carex elata* 'Aurea' in the foreground.

Border basics

Size 3x2m (10x6ft)

Suits Formal and informal gardens, as a screen or divide between sections

Soil Free-draining to moist

Site Full sun to partial shade

Shopping list

- 3 x *Cortaderia selloana* 'Sunningdale Silver'
- 3 x *Dahlia* 'Bishop of Llandaff'
- 2 x *Sedum* 'Herbstfreude'
- 2 x *Carex elata* 'Aurea'

Planting and aftercare

Space the pampas grasses along the line of the screen or hedge, with approximately 1m (3ft) between them or, for immediate cover, use more plants and place them closer together.

Plant the dahlias in a group at least 60cm (2ft) in front of the pampas grass – if you garden in a cold area, lift the dahlias in autumn and move them to a frost-free place over winter. Group the sedums and the sedges together in the foreground.

Cortaderia selloana 'Sunningdale Silver' ❋❋❋ ◊ ◑ ☼ ☀ ♈

Dahlia 'Bishop of Llandaff' ❋❋ ◑ ☼ ☀ ♈

Sedum 'Herbstfreude' ❋❋❋ ◊ ◑ ☼ ☀ ♈

Carex elata 'Aurea' ❋❋❋ ◊ ◑ ☼ ☀ ♈

Bamboo camouflage

A lovely planting that combines bamboos, grasses, and sedges extremely well but also serves a practical purpose, screening a garden shed from view.

Fargesia murielae, with its elegant arching shape, and more upright *Phyllostachys aureosulcata* f. *aureocaulis* intermingle at the heart of the border. Their lower branches have been pruned to allow for the underplanting of grasses and drought-tolerant New Zealand sedges in a range of contrasting but complementary colours.

Border basics

Size 4x2m (12x6ft)

Suits Informal, contemporary, and oriental gardens

Soil Free-draining

Site Full sun to partial shade

Shopping list:

- 1 x *Fargesia murielae*
- 1 x *Phyllostachys aureosulcata* f. *aureocaulis*
- 1 x *Pennisetum alopecuroides* 'Hameln'
- 1 x *Leymus arenarius*
- 2 x *Carex buchananii*
- 3 x *Carex comans*

Planting and aftercare

Plant the two bamboos centrally and quite close together so that they provide the best possible screening effect. Place the grasses and sedges in their pots around the bamboos, with the taller *Carex buchananii* and *Leymus* closest to them and smaller plants towards the edges of the bed. When happy with their positions, plant each in turn, working from the middle outwards. This planting will largely look after itself. Simply trim back the deciduous grasses in spring, remove dead leaves from the *Carex*, and keep the bamboos from wandering too far.

Fargesia murielae
❋❋❋ ◊ ◗ ☼ ☀ ⚲

Phyllostachys aureosulcata f. *aureocaulis* ❋❋❋ ◊ ◗ ☼ ☀ ⚲

Pennisetum alopecuroides 'Hameln'
❋❋❋ ◊ ◗ ☼

Leymus arenarius
❋❋❋ ◊ ☼ ☀

Carex buchananii
❋❋/❋❋❋ ◊ ◗ ☼ ☀ ⚲

Carex comans
❋❋/❋❋❋ ◊ ◗ ☼ ☀

Elegant summer show

A sophisticated, evocative combination of only four plants, this grouping nevertheless provides a long season of interest and contrasts, peaking at the height of summer. By then the *Allium* flowers will have already faded, but they still retain an important structural role, introducing contrasting spherical forms to the upright green daggers of *Crocosmia* foliage and its startlingly vivid red flowers. The billowing blooms of *Stipa gigantea* and *Stipa tenuissima* frame the *Crocosmia* above and below, and will last long after the *Crocosmia* has faded.

Border basics

Size 2x2m (6x6ft)

Suits Informal, contemporary, and prairie-style gardens

Soil Free-draining

Site Full sun

Shopping list

- 1 x *Stipa gigantea*
- 2 x *Crocosmia* 'Lucifer'
- 6 x *Allium* bulbs, such as
  A. giganteum
- 2 x *Stipa tenuissima*

Planting and aftercare

Plant the *Stipa giganteum* at the centre of this composition, with the *Crocosmia* and *Stipa tenuissima* in front and to either side. All these plants are fairly vigorous, and will tolerate being placed quite closely together.

For the most naturalistic effect, scatter the *Allium* bulbs in front of the *Crocosmia* and plant them where they land. Their purple spheres will provide the early colour in this scheme, harmonizing with the fresh green foliage of the other plants in late spring.

Stipa gigantea
❋❋❋ ◊ ☼ ♈

Crocosmia 'Lucifer'
❋❋❋ ◊◗ ☼ ◑ ♈

Allium
❋❋❋ ◊◗ ☼ ◑

Stipa tenuissima
❋❋❋ ◊◗ ☼ ◑

Artistic inspirations

In this painterly composition, grasses and sedges add movement and texture to a herbaceous underplanting that repeats the same few plants.

Stipa gigantea is used here so that its height is emphasized, but the transparent nature of its flowerheads means it never dominates. Spherical *Allium* flowers also add vertical notes, while the yellow, strap-shaped leaves of golden sedge (*Carex elata* 'Aurea') contrast with the leaf textures and colours of *Astrantia*, purple-leaved *Persicaria*, and steely blue *Perovskia*.

Border basics

Size 4x4m (12x12ft)

Suits Informal and contemporary gardens

Soil Moist but free-draining

Site Full sun to part shade

Shopping list

- 2 x *Stipa gigantea*
- 3 x *Perovskia* 'Blue Spire'
- 2 x *Persicaria microcephala* 'Red Dragon'
- 8–12 x *Allium giganteum*
- 5 x *Astrantia* 'Hadspen Blood'
- 3 x *Carex elata* 'Aurea'

Planting and aftercare

This scheme relies on closely packed plants for impact. Plant the *Stipa* first, backed by the *Perovskia*, with the *Persicaria* in front of that. Group the same plants together, but weave the groupings through each other to create a naturalistic feel, with the *Carex* and *Astrantia* towards the front.

The plants are so close together that you will probably need to cut back the more vigorous species, such as the *Perovskia* and *Persicaria*. Extra water may be required in dry periods. Deadheading will extend the flowering period, and is well worth the extra effort.

Stipa gigantea
❀❀❀ ◊ ☼ ♈

Perovskia 'Blue Spire'
❀❀❀ ◊ ☼ ◑ ♈

Persicaria microcephala 'Red Dragon'
❀❀❀ ◊ ◐ ☼ ◑

Allium giganteum
❀❀❀ ◊ ◐ ☼ ◑ ♈

Astrantia 'Hadspen Blood'
❀❀❀ ◊ ◐ ☼ ◑

Carex elata 'Aurea'
❀❀❀ ◐ ◑ ◑ ◑ ♈

Creative containers

This simple but really elegant grouping comprising a sedge and two grasses in pots in front of a bamboo screen would grace almost any style of garden.

The bamboo *Fargesia nitida* forms an evergreen backdrop, with Bowles' golden sedge (*Carex elata* 'Aurea') cascading out of the largest pot, and steely blue *Elymus magellanicus* and arching clumps of variegated moor grass (*Molinia caerulea*) in front. The main impact comes from the contrast in leaf colours.

Border basics

Size 1x1m (3x3ft) corner of a patio or decking

Suits Contemporary, informal, and oriental gardens

Soil Potting compost

Site Full sun to partial shade

Shopping list

- 2 x *Fargesia nitida*
- 1 x *Carex elata* 'Aurea'
- 1 x *Molinia caerulea* subsp. *caerulea* 'Variegata'
- 1 x *Elymus magellanicus*

Planting and aftercare

Plant the two bamboos in the ground where they will create the backdrop for your group of pots. Choose three pots in the same material but in different sizes, the largest being about 30cm (12in) in diameter. Glazed oriental-style or galvanized metal containers would serve just as well as terracotta. Half-fill each pot with a loam-based, John Innes-type compost. Plant the golden sedge in the largest pot, *Elymus* in the medium pot, and moor grass in the smallest. Arrange the three until you are happy with the grouping. All three plants are deciduous, dying back in winter. The *Elymus* is not reliably hardy, so may need extra protection in winter in colder areas.

Fargesia nitida
❀❀❀ ◊ ◖ ☼ ☀

Carex elata 'Aurea'
❀❀❀ ◊ ◖ ☼ ☀ ♉

Molinia caerulea subsp. *caerulea* 'Variegata' ❀❀❀ ◊ ◖ ☼ ♉

Elymus magellanicus
❀❀/❀❀❀ ◊ ◖ ☼

Flower and grass pots

Combining grasses with tender annual summer bedding plants produces very interesting and attractive displays, which can be varied from year to year.

In this understated but no less lovely container planting, the feathery flowers of *Stipa tenuissima* are a foil to burgundy red *Verbena* 'Diamond Merci' flowers, while the small white flowers of *Sutera cordata* 'Snowflake' provide a cascading foreground. Many other grasses are suitable for use in mixed containers, such as *Pennisetum villosum* or *Deschampsia*.

Container basics

Size Large pot at least 25cm (10in) in diameter

Suits Cottage, contemporary, and informal gardens

Soil John Innes or multi-purpose compost

Site Full sun to partial shade

Shopping list

- 1 x *Stipa tenuissima* or *Pennisetum villosum*
- 3 x *Verbena* 'Diamond Merci'
- 3 x *Sutera cordata* 'Snowflake'

Planting and aftercare

The perennial *Stipa* can be planted permanently in a large pot, with new annual partners added each year in early summer, once all risk of frost has passed. Plant the grass towards the back of the container, leaving plenty of space in front and to the sides for the other plants. Plant the *Verbena* in front of the grass, leaving room for the *Sutera cordata* to be fitted in close to the edges where they can trail over the sides. Feed and water the container regularly during the summer months, and deadhead the *Verbena* plants to keep them flowering strongly. Remove the annuals after they die back in autumn.

Stipa tenuissima
❄❄❄ ◊ ◊ ☼ ☼

Verbena 'Diamond Merci'
❀◊ ◊ ☼ ☼

Sutera cordata 'Snowflake' (formerly *Bacopa cordata*) ❀◊ ◊ ☼ ☼

Alternative plant idea

Pennisetum villosum
❄❄ ◊ ◊ ☼ ♈

Tropical mix

The four substantial plants in this grouping are all quite hardy yet look very tropical. This flowerless composition relies on contrasts in plant form, foliage textures, and colouring for effect.

The stout trunk of the tree fern (*Dicksonia antarctica*) complements the upright *Phormium* and *Miscanthus*. The fern is topped by a contrasting, arching canopy of fronds, subtly echoed in the spreading shape of the bamboo (*Phyllostachys nigra*).

Border basics

Size 4x3m (12x10ft)

Suits Contemporary, subtropical, and informal gardens

Soil Moisture-retentive

Site Partial shade

Shopping list

- 1 x *Dicksonia antarctica*
- 1 x *Phyllostachys nigra*
- 1 x *Phormium tenax* 'Variegatum'
- 1 x *Miscanthus sinensis* 'Zebrinus'

Planting and aftercare

These are large plants when mature, so space them widely apart. Plant the *Dicksonia* at the back of the border, inclining it slightly outwards, if desired. Plant the bamboo to the rear of the opposite corner of the border, at least 2m (6ft) from the tree fern. Position the *Phormium* in front of the tree fern, and the *Miscanthus* between the tree fern and the bamboo.

This is a relatively low-maintenance planting, although the tree fern needs watering in dry periods – water the crown, where the fronds originate, not the base. *Dicksonia* is not reliably hardy in very cold regions, so protect it in winter by wrapping the stem and crown in fleece or covering them with straw held in place with chicken wire.

Dicksonia antarctica
❄❄/❄❄❄❄ ◊ ☼ ☼ ☽ ♈

Phyllostachys nigra
❄❄❄ ◊◊ ☼ ☽ ♈

Phormium tenax 'Variegatum'
❄❄❄ ◊◊ ☼ ☽ ♈

Miscanthus sinensis 'Zebrinus'
❄❄❄ ◊◊ ☼ ♈

Jungle effects

Statuesque bamboos are underplanted here with shade-tolerant sedges and woodrushes, giving this composition a lush, jungly feel. This planting would look great in a subtropical-style garden with larger-leaved plants, but would also perform well in the corner of a town garden shaded by buildings or large trees.

As they mature, the bamboos will knit together, forming a screen. The sedges and woodrushes will also grow to fill in the gaps, giving continuous, weed-suppressing ground cover.

Border basics

Size 3x1.5m (10x5ft), against a wall or to create a divide within the garden
Suits Contemporary, subtropical, and oriental styles of garden
Soil Free-draining to moist
Site Partial to quite deep shade

Shopping list

- 1 x *Phyllostachys bambusoides* or *Fargesia murielae*
- 1 x *Phyllostachys nigra*
- 3 x *Carex oshimensis* 'Evergold'
- 2 x *Luzula nivea*
- 3 x *Luzula sylvatica*

Planting and aftercare

Plant the bamboos (*Phyllostachys*) first, towards the back of the border, at least 1m (3ft) apart to give them room to spread. Place the woodrushes (*Luzula*) to either side and group the variegated sedges (*Carex*) in the centre. These plants also spread, so space them out evenly. When you are happy with the arrangement, go ahead and plant. Cover with a mulch of chipped bark to keep weeds down. Bamboos and woodrushes are drought-tolerant once established, but the sedge may need watering during dry periods. Woodrushes may spread rapidly, so keep them in check.

Phyllostachys bambusoides
❀❀❀ ◌ ◍ ☼ ☀

Phyllostachys nigra
❀❀❀ ◌ ◍ ☼ ☀ ♔

Carex oshimensis 'Evergold'
❀❀❀ ◌ ◍ ☼ ☀ ♔

Luzula nivea
❀❀❀ ◌ ◍ ☀ ☀

Luzula sylvatica
❀❀❀ ◌ ◍ ☀ ☀

Alternative plant idea

Fargesia murielae
❀❀❀ ◌ ◍ ☼ ☀ ♔

Naturalistic bog garden

This small-scale, green-leaved planting looks naturalistic and subtle around a tiny pond or section of stream. The heart-shaped leaves of *Houttuynia cordata* are seen at the front of the planting as well as in the background – repeating plants in this way helps tie a scheme together visually. The flamboyant variegated *Houttuynia cordata* 'Chameleon' is more widely available, but it would have dominated here. Instead, the broad leaves of the plain species contrast well with the upright, narrow stems of the horsetail (*Equisetum*), while the even broader leaves of the arum lily (*Zantedeschia*), *Hosta,* and *Petasites* complete the picture.

Zantedeschia aethiopica 'Crowborough'
❋❋/❋❋❋❋ ◐● ☼ ☀ ♉

Hosta sieboldiana
❋❋❋ ◐ ☼ ☀

Border basics

Size 2x2m (6x6ft), around a pond or stream

Suits Bog and water gardens, wildlife gardens

Soil Moist to wet

Site Full sun to part shade

Shopping list

- 1 x *Zantedeschia aethiopica* 'Crowborough'
- 1 x *Hosta sieboldiana*
- 1 x *Equisetum hyemale* var. *affine*
- 1 x *Carex glauca*
- 1 x *Petasites japonicus*
- 3 x *Houttuynia cordata*

Equisetum hyemale var. *affine*
❋❋❋ ◐● ☼ ☀

Carex glauca
❋❋❋ ◐● ☼ ☀

Planting and aftercare

The horsetail can be grown in the water in aquatic planting baskets (*see pp.56–57*). Place the remaining plants towards the water's edge, the *Petasites* front left, one *Houttuynia* and the *Carex* front right. Plant the other two *Houttuynia*, the *Zantedeschia* (which could also be grown in water like the horsetail), and the *Hosta* at the back.

Pond edges can dry out and pond water levels fall in hot weather, so keep both topped up with water.

Petasites japonicus
❋❋❋ ◐● ☼ ☀

Houttuynia cordata
❋❋❋ ◐● ☼ ☀

Modern poolside planting

Uncompromisingly contemporary, this pond planting may not appeal to everyone, but makes marvellous use of foliage colours and a mirrored screen to create a dark, somewhat brooding look.

The two yellow full-moon maples (*Acer shirasawanum* 'Aureum'), blocks of repeating hair grasses (*Deschampsia*), and the purple-leaved *Heuchera* form the backbone of the composition. The *Heuchera* and fescues (*Festuca glauca*) pick up the blue tinges of the pebbles on the bed of the pond and its edging.

Border basics

Size 4x3m (12x10ft), around a pond

Suits Contemporary gardens

Soil Moisture-retentive

Site Full sun to part-shade

Shopping list

- 4 x *Deschampsia flexuosa* 'Tatra Gold'
- 6 x *Deschampsia cespitosa* 'Goldtau'
- 2 x *Acer shirasawanum* 'Aureum'
- 4 x *Heuchera micrantha* var. *diversifolia* 'Palace Purple'
- 3 x *Festuca glauca* 'Blaufuchs'
- 2 x *Hosta* 'August Moon'

Planting and aftercare

Plant the four *Heuchera* at the water's edge towards the back, with the six *Deschampsia cespitosa* 'Goldtau' evenly spaced behind them, and the yellow-leaved 'Tatra Gold' group around the corner of the pond. The foreground is a medley of contrasting foliage, with alternating plants of *Acer* and *Hosta* near the pond edge, and blue fescue grasses behind and between them. Finish off with a decorative mulch, such as glass chippings or rounded pebbles.

Water the plants well while they become established and in dry spells, and top up water in the pool. Cut back the *Deschampsia* in late autumn.

Deschampsia flexuosa 'Tatra Gold'
❄❄❄ ◊ ☼ ☼

Deschampsia cespitosa 'Goldtau'
❄❄❄ ◊ ☼ ☼

Acer shirasawanum 'Aureum'
❄❄❄ ◊ ☼ ☼ ♛

Heuchera micrantha var. *diversifolia* 'Palace Purple' ❄❄❄ ◊ ◊ ☼ ☼

Festuca glauca 'Blaufuchs'
❄❄❄ ◊ ◊ ☼ ☼ ♛

Hosta 'August Moon'
❄❄❄ ◊ ☼ ☼

Looking after your plants

This chapter is a guide to keeping your bamboos, grasses, and grass-like plants looking their best. Most of the plants in this book are easy to maintain, but at certain times of the year, especially in spring and autumn, a little care and attention pays dividends. You may need to keep spreading bamboos in check, and prune out old grass and bamboo stems to allow space for new growth to push through. Plants that have outgrown their containers will also need to be repotted.

Caring for bamboos

Once established, most bamboos and grasses are fairly drought-tolerant and need watering only in very dry spells. Newly planted and establishing plants, however, should be kept moist during their first summer.

Feeding with mulches Mulches of organic matter, such as compost, composted bark (*right*), or well-rotted manure, conserve water in the soil and gradually release nutrients as they break down. A 5–7cm (2–3in) layer of composted bark will feed a bamboo throughout the year. Dead bamboo leaves (*below*) are an excellent mulch for bamboos because they provide plants with silica, which they need to form wood.

Watering methods The simplest way of watering is with a can (*below left*). This allows you to target water at the plant's root ball and not the foliage. It is better to water thoroughly once or twice a week than little and often, which encourages roots to grow towards the surface and makes plants more vulnerable during drought. Automated systems take the effort out of watering. Drip nozzles (*below centre*) can be placed exactly where they're needed, while seep hoses (*below right*) placed on the soil surface under a mulch water most efficiently.

Deliver water right to the base of the plant and soak the roots thoroughly.

Drip nozzles are an effective irrigation method and can also be used in containers.

Seep hoses irrigate efficiently and are good for beds, borders, and boggy areas.

Feeding container-grown plants
Container-grown plants have a restricted root run and limited compost from which they can absorb nutrients, making feeding an important task. Give a regular liquid feed when watering or sprinkle slow-release fertilizer granules, which release nutrients slowly over a year or so, onto the surface of the compost. Alternatively, incorporate granules into the compost when repotting the plant. Plants in pots also need watering more frequently than those in the ground whose roots can search deep below the surface for moisture.

Caring for grasses

Grasses and grass-like plants are among the easiest of all garden plants to look after, but low-maintenance is not no-maintenance: a little regular attention will keep them looking their best.

Preventing excess seedlings Some grasses and grass-like plants, such as this *Luzula nivea*, disperse a lot of viable seed, resulting in unwanted plants. To avoid this, simply remove any fading flowers with scissors or secateurs before they start shedding seed.

Identifying signs of stress

Healthy grass leaves are usually flat and green all the way to the tips (*near right*), indicating that the plant is growing well. During prolonged dry periods the leaves of many grass species roll up into tubes (*far right*), showing that the plant is stressed. This reaction actually helps to minimize moisture loss by reducing the exposed surface area of the leaves, making them less vulnerable to the drying effects of the sun and wind. If conditions are wet, particularly after a rainy winter, and the rolling is accompanied by browning or yellowing, it may be that the roots are too wet. In this case, move the plant to a drier position.

Healthy grass with flat, green leaves.   Stressed grass with leaves rolled into tubes.

Cutting out dead or diseased foliage Even the leaves of fully evergreen grasses and grass-like plants have a finite lifespan, and usually begin to die off from the tips downwards. Trim off yellow or brown areas with scissors, or cut the whole leaf away at its base.

Spring cleaning Older leaves may go brown in winter. Affected leaves of small plants like some sedges, *Acorus*, *Luzula*, and *Ophiopogon* can be removed individually, but with larger plants like this *Festuca*, it is quicker to comb the clump with a spring-tined rake.

Cutting back deciduous grasses

It is important to cut deciduous grasses back hard if you want to keep them looking their best. You can either do this in late autumn or, if you want your grasses to provide winter structure and interest in the garden, early spring.

Tip for success

Where the dead stems of a clump-forming deciduous grass have collapsed, remove them with a spring-tined rake.

1 In late winter or early spring, the previous summer's browned stems of large grasses, such as *Miscanthus* or *Stipa*, should be cut back to near ground level with secateurs. Try not to damage any new green shoots.

2 Work your way methodically through the clump, removing all the old stems (which can be composted). Some grasses can be divided once cut back, but *Miscanthus* is best split when in active growth.

Alternative methods

The thick stems or sharp-edged leaves of larger *Miscanthus*, or the flower stems of *Cortaderia selloana* (pampas grass), can be too tough for garden secateurs (and the gardener's bare skin). Protect your hands by wearing gloves and use loppers for cutting.

If you have a large number of grasses to get through, it will be quicker to use an electric hedge trimmer. Cut the stems at a slightly higher level than you would with secateurs to avoid damaging any new growth emerging at the base.

Pruning and tidying bamboos

Gardeners have a tendency to be afraid of pruning bamboos, but they shouldn't be. Most species respond really well to being cut back, and it can, in fact, increase their vigour.

Thinning out an established clump Clumps of bamboo benefit greatly from being thinned every one to two years in early spring, before they begin shooting. Cut any dead (brown and leafless), thin, or weak culms (stems) down to ground level using sharp secateurs. Removing a quarter to a third of the culms is usually enough but, if the bamboo is congested, don't be afraid to take out up to half of the clump, or even more if necessary, as is the case here. This opens up the clump, reinvigorating it and giving space for new culms to grow within the clump rather than around its edges.

Retaining moisture All the dead brown culms have been removed from this neglected clump, leaving only the strong green new growth. To aid moisture retention and for a more decorative effect, apply a gravel mulch. New culms will have no problem pushing through the gravel.

Reducing height Bamboo culms tend to get thicker and taller as the plant matures. If a bamboo has grown too tall, shorten each culm to the desired height, cutting just above a node with branches. If a bamboo is too leafy, cut out some of the branches to lighten the canopy.

Picking off dead leaves It is a characteristic of most bamboos to hold on to yellow or dead brown leaves and not shed them in the way that deciduous plants do. These leaves can be picked off individually from smaller plants to make them look greener, but it is a tedious process with large plants, so it may be better to accept them as part of their charm. Don't throw away old leaves; use them as a silica-rich mulch under the plants.

Finishing touches The papery sheaths that protect new bamboo culms persist in many species, and can look quite ornamental, giving the plants a variegated look. But some plants, such as the black bamboo *Phyllostachys nigra*, many other *Phyllostachys*, and *Thamnocalamus* species, have attractively coloured stems that are best shown off by removing the sheaths by hand (*far left*). The stems of such species can be further enhanced by polishing them with a soft cloth (*left*) to remove any powdery "bloom". To really make them shine, add a few drops of vegetable oil to the cloth.

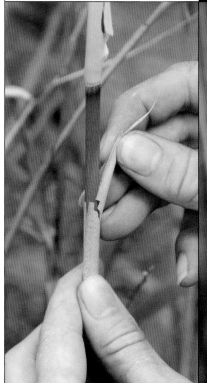

Remove papery sheaths to show the culms.

Polishing adds shine to bamboos.

Controlling a bamboo's spread

Once bamboos have established, which may take several years, new stems will appear around the edges. These can be snapped off when short. Alternatively, an annual "chop" around the root ball in spring will curb their vigour.

Tip for success

A sharp spade makes it easier to cut around the root ball. Use a sharpening stone to maintain a good chopping edge.

1 The simplest way of controlling a clump-forming bamboo is to prune around the edges of its root ball, removing the rhizomes (underground stems) that produce the new culms. Dig a trench 30cm (12in) deep around the clump.

2 If you encounter any tough rhizomes, either sever them with a sharpened spade or cut with secateurs where they emerge into the trench.

3 Either pull cut rhizomes out of the ground, or uncover them using a fork and remove. Destroy the cut sections. Alternatively, use them to propagate new bamboo plants by simply potting them up.

4 Refill the trench with the excavated soil. Any new culms that appear can be snapped off at ground level; they will not regrow. Thinning out the oldest, weakest culms will encourage the bamboo to reshoot within the clump.

Maintaining container-grown bamboos

Container-grown bamboos lose vigour as the pots fill with roots, and must be split and repotted every two or three years to keep them looking good. Large, very vigorous plants may need annual attention, ideally in early spring.

1 Remove the plant from its pot – this picture shows just how congested the root system of a bamboo can become in only a few years. There is little space for compost and, therefore, little water-holding capacity.

2 Using a stout saw (or even an axe), which is necessary to cut through the tough rhizomes, take off the lower quarter, or at least 3–4cm (1¼–1½in), of the root ball and discard it.

3 Look for a natural line or gap through the plants' top growth and saw down the line, dividing the plant into halves, thirds, or quarters, depending on its size.

4 Ensure that each section has a reasonable root system, with at least half a dozen vigorous stems (any dead or weak ones can be cut out at this stage), and, ideally, several new shoots visible near the surface.

Maintaining container-grown bamboos *continued*

5   Part of the old plant can be grown in the old pot or a new, larger one; other pieces can be planted in the garden. Add plenty of crocks to the base of the pot for drainage, and several centimetres of loam-based compost.

6   Position a section in the pot to gauge the planting depth. Place the curved, outside part of the root ball in the middle. This puts the most vigorously growing part of the plant where there is most fresh compost.

7   Fill around the division with fresh compost and firm the soil down gently with your fingers. Ensure that the divided section remains upright for a neat appearance.

8   Top up the compost so that the surface of the old root ball is covered with 2–3cm (¾–1¼in) of new soil – bamboos can "heave" themselves up in a pot over time. Work the compost between the culms for an even surface.

9 Water the plant well to settle the compost around the stems. If the container is large, it is usually prudent to place it in its intended position before this stage because wet compost is considerably heavier than dry.

10 Cover the soil with a moisture-retaining mulch of bark chippings, gravel, or more decorative water-rounded pebbles, as shown here.

11 The iridescent modern pot complements the bamboo's architectural qualities well. Although it can take a while for a divided bamboo to recover its vigour, once it has, it will grow away strongly and soon fill the pot.

Seasonal planner: spring and summer

This planner outlines the jobs you should be doing at different times of the year. Follow these simple guidelines to ensure that your bamboos, grasses, sedges, and other grass-like plants, remain healthy, strong, and looking good.

Spring

Bamboos

- Early spring is a good time for planting container-grown bamboos, before growth has started in earnest.
- Thin out dead, old, and weak stems from established clumps of bamboo to encourage new growth.
- Snap off emerging shoots coming up in the wrong place at ground level – this causes them to abort, so they will not regrow.
- Remove yellowing leaves to keep plants looking groomed – a percentage of old leaves will yellow over the course of the winter, but this is no cause for concern.
- Top-dress clumps with a mulch or garden compost.

Grasses

- Sow seed of annual and perennial grasses in trays or modules (*below*).
- Give clumps of deciduous grasses their annual "haircut", ideally just before or as new green growth starts to appear, if they were not cut back in autumn.
- Comb the dead leaves out of clumps of evergreen grasses, using a rake for larger species or fingers for smaller ones (wear gloves if the leaf edges are sharp).

Sedges and other grass-like plants

- Comb dead leaves out of evergreen clumps of plants, and cut off dead leaf tips. Deciduous species should have been cut back in autumn.

Summer

Bamboos

- Water newly planted bamboos during dry periods while they are establishing.
- Water container-grown bamboos regularly, never allowing them to dry out completely, which can result in all the leaves browning.
- To show off the stem colour of an established clump, prune off the lowest branches close to the stems.
- Take off persistent leaf sheaths from stems once they begin leafing out (unless you like the "variegated" effect they produce).

Grasses

- It is still not too late to plant container-grown grasses (*below*); there should be a better selection in garden centres than earlier in the year.
- Water newly planted grasses in prolonged dry periods, particularly those in pots.
- Late summer is the best time to divide *Miscanthus* species and cultivars, but not other grasses.

Sedges and other grass-like plants

- Keep moisture-lovers like most sedges, rushes, horsetails, and reeds well watered. The soil around artificial, lined ponds can dry out almost as quickly as elsewhere in the garden, and may need watering once or twice a week to keep these plants happy.

Sow annual or perennial grasses in modules in spring.

Container-grown grasses can still be planted out in summer.

Seasonal planner: autumn and winter

Autumn

Bamboos
- Autumn is a good time to thin bamboo clumps (*below*), removing thin and weak culms, those that have died over summer, or those that are too far from the clump.
- Top-dress with an organic mulch or apply a slow-release general-purpose fertilizer.
- Continue watering potted bamboos as and when necessary – autumn can be as warm and dry as summer.
- If a bamboo clump has spread too far, dig a trench around it and cut off the rhizomes (*see pp.110–111*).

Grasses
- Deadhead deciduous grasses that can self-seed too enthusiastically, such as *Molinia* and *Festuca* species, removing the seedheads before they ripen.
- Most grass species, excluding *Miscanthus*, can be divided successfully in autumn.
- Continue watering grasses grown in containers.

Sedges and other grass-like plants
- Cut back deciduous water plants, such as *Phragmites* and *Typha,* as the leaves and stems go brown. Remove the dead material from the water; it will only rot.
- Divide sedges, horsetails, and plants grown in aquatic baskets, potting up or planting out the divisions.
- Keep watering pot-grown moisture lovers like sedges in mild, dry periods.

Winter

Bamboos
- Late autumn and early winter are the best times to plant bare-root bamboo divisions (*see pp.50–51*).
- Divide clumps of bamboos growing in the ground, especially if they are too large (*see pp.110-111*).
- In mild periods, winter is the best time to split and repot container-grown bamboos (*see pp.112–115*).
- Check if container-grown bamboos are moist enough, watering as necessary.

Grasses
- Small deciduous grasses can collapse in winter, the dead leaves smothering their crowns, keeping them too damp. Cut back and remove the leaves when necessary.
- Larger deciduous species, such as *Miscanthus*, can be left over winter (*below*), but cut back stems if they collapse.
- Cut back fading flower stems on evergreen grasses, such as *Cortaderia* (pampas grass) and *Stipa gigantea*.

Sedges and other grass-like plants
- Continue cutting back deciduous species as their leaves go brown, removing and composting the material.
- Deadhead evergreen species as their flower stems fade and die off.
- Late winter and early spring are good times to divide sedges, both evergreen and deciduous, and plants such as *Acorus* (sweet rush) and *Ophiopogon* (mondo grass).

Use loppers to thin out overgrown clumps of bamboo.

Frosted *Miscanthus* can add interest to the winter garden.

Plant guide

All the bamboos and grasses profiled here are reliably hardy, with one or two exceptions, and although most garden centres stock only a limited range, you will find more at specialist nurseries (*see pp.154–55*). The symbols below are used throughout the guide to indicate the conditions each plant prefers.

Key to plant symbols

 ♀ Plants given the RHS Award
    of Garden Merit

Soil preference

 ◊ Well-drained soil
 ◑ Moist soil
 ● Wet soil

Preference for sun or shade

 ☼ Full sun
 ☼ Partial or dappled shade
 ☀ Full shade

Hardiness ratings

 ✻✻✻ Fully hardy plants
 ✻✻ Plants that survive outside in mild
     regions or sheltered sites
 ✻   Plants that need protection from frost
     over winter
 ❀   Tender plants that do not tolerate any
     degree of frost

Bamboos (Chusquea–Indocalamus)

Chusquea culeou
This unusual South American bamboo forms dense clumps of upright, golden yellow culms. With more than 50 short branches growing upright from each node, it has a distinctive "bottlebrush" appearance. It spreads relatively slowly.

H: to 6m (20ft), **S**: 2.5m (8ft)
✻✻✻ ◊ ◐ ☼ ◑ ♈

Fargesia murielae
Fargesia species and cultivars are extremely hardy and cope well with exposed sites and dry soils. *Fargesia murielae* is one of the toughest, with light green culms, an arching habit, and a slow rate of spread. Its leaves are about 8cm (3in) long.

H: to 4m (12ft), **S**: 4m (12ft)
✻✻✻ ◊ ◐ ☼ ◑ ♈

Fargesia murielae 'Simba'
A dwarf and exceptionally leafy *Fargesia* about half the size of the species, 'Simba' forms a semi-weeping fountain of apple green leaves, and does particularly well in containers. Its culms and branches are an attractive orange-red colour.

H: to 2m (6ft), **S**: 2m (6ft)
✻✻✻ ◊ ◐ ☼ ◑ ♈

Fargesia nitida

Similar to *F. murielae*, but more upright and even more tolerant of exposed sites, *F. nitida* makes a good windbreak or hedge. Newly produced culms often do not branch until their second year, and may be attractively purple-flushed.

H: to 4m (12ft), **S**: 4m (12ft)
❀❀❀ ◌ ◑ ☼ ☀

Fargesia robusta

A relatively recent introduction with upright growth and large leaves up to 13cm (5in) long. Its dark green culms have prominent white sheaths when they emerge, making the stems look variegated. It is as hardy and wind resistant as most *Fargesia*.

H: to 4m (12ft), **S**: 4m (12ft)
❀❀❀ ◌ ◑ ☼ ☀

Himalayacalamus falconeri

This large, handsome plant from the Himalayas is not the hardiest of bamboos, preferring some shelter. It forms tight clumps and in most gardens reaches only around 4m (12ft) – barely half its height in the wild. It does well in containers.

H: to 7m (22ft), **S**: 4m (12ft)
❀❀/❀❀❀ ◌ ◑ ☼ ☀

Indocalamus tesselatus

Indocalamus have the largest leaves of any temperate bamboo; those of *I. tesselatus* are up to 40cm (16in) long. The plant forms a dome of foliage about 1m (3ft) high. It does well in a container and looks very tropical, though it is fully hardy.

H: to 1.7m (5.5ft), **S**: indefinite
❀❀❀ ◌ ◑ ☼ ☀ ♛

Bamboos (Phyllostachys)

Phyllostachys aureosulcata f. aureocaulis
A large bamboo with rich golden yellow culms when it is grown in sun. The culms are also unusual in that some of them zig-zag crookedly lower down. The plant usually grows no more than 8m (25ft) in height.

H: to 8m (25ft), **S**: 4m (12ft)
✻✻✻ ◌ ◗ ☼ ☀ ♆

Phyllostachys aureosulcata f. spectabilis
Similar to the preceding plant, although its culms, which have green stripes in the grooves, are an even richer gold. The grooves characterize the genus, as do a clumping habit and relatively slow spread.

H: to 8m (25ft), **S**: 4m (12ft)
✻✻✻ ◌ ◗ ☼ ☀ ♆

Phyllostachys bambusoides
A giant bamboo used for its timber in the Far East, *P. bambusoides* can be tree-like, but grows much smaller in temperate gardens, really preferring warmer summers. Given space, its thick green culms and 15cm (6in) leaves are magnificent.

H: to 22m (74ft), **S**: 6m (20ft)
✻✻✻ ◌ ◗ ☼ ☀

Phyllostachys flexuosa
Another giant bamboo, *P. flexuosa*, as its name suggests, has flexible, shiny green culms that tend to arch over, giving the plant an attractive weeping form. The culms of some plants may age to an attractive mottled yellow-black.

H: to 10m (30ft), **S**: 6m (20ft)
✻✻✻ ◌ ◗ ☼ ☀

Phyllostachys nigra
The most popular temperate bamboo, and rightly so. The culms emerge olive green but turn glossy black with age and in sun, contrasting beautifully with the 10cm (4in) fresh green leaves. Usually 3–5m (10–15ft) tall, it is slow-growing and stays compact.

H: to 15m (50ft), **S**: 6m (20ft)
❀❀❀ ◊ ◐ ☼ ☼ ♉

Phyllostachys nigra f. henonis
This form is similar to *P. nigra* in size and vigour, but the culms are brownish-green and do not turn black. This may actually be the wild form of black bamboo, from which the cultivated, black-stemmed plants were selected.

H: to 15m (50ft), **S**: 6m (20ft)
❀❀❀ ◊ ◐ ☼ ☼ ♉

Phyllostachys violascens
One of the most striking species of *Phyllostachys*. The culms emerge deep violet and turn brown with irregular brownish-purple stripes as they age. You may need to curb this plant's vigour by cutting out the old culms as new ones begin to leaf.

H: to 11m (35ft), **S**: 4m (12ft)
❀❀❀ ◊ ◐ ☼ ☼

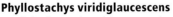

Phyllostachys viridiglaucescens
A tall, leafy species with green culms and fairly large leaves that makes an effective windbreak, since it spreads relatively quickly once established. There are, however, other bamboos suited to the task that have more ornamental features.

H: to 10m (33ft), **S**: 6m (20ft)
❀❀❀ ◊ ◐ ☼ ☼

Phyllostachys vivax f. aureocaulis
This species is broadly similar to *P. bambusoides*, with thinner culms and more arching foliage. It copes better with cooler climates, not reaching anywhere near the 15m (50ft) it attains in the wild. The culms are a beautiful golden yellow.

H: to 8m (25ft), **S**: 4m (12ft)
❀❀❀ ◊ ◐ ☼ ☼ ♉

Bamboos (Pleioblastus–Semiarundinaria)

Pleioblastus pygmaeus
All species of *Pleioblastus* are rampant and grow well in shade. Despite its diminutive size, this pretty little bamboo can be invasive in the garden. Nevertheless it makes good ground cover under trees, and grows well in a container.

H: to 60cm (24in), **S**: 4m (12ft)
❋❋❋ ◊ ◑ ☼ ☀

Pleioblastus variegatus
This is one of the best variegated bamboos, with broad irregular cream striping on each leaf. It forms dense short groves, usually much less than 2m (6ft) tall. It may spread, but slowly for a *Pleioblastus*, and also grows well in wide, shallow containers.

H: to 2m (6ft), **S**: 4m (12ft)
❋❋❋ ◊ ◑ ☼ ☀ ♉

Pseudosasa japonica
Pseudosasa are rampant bamboos, but they spread more slowly than *Sasa* or *Indocalamus*. The large leaves may brown in cold windy weather, so although fully hardy, *P. japonica* likes a sheltered position. It can make an attractive boundary screen.

H: to 4m (12ft), **S**: 4m (12ft)
❋❋❋ ◊ ◑ ☼ ☀ ♉

Pleioblastus viridistriatus
(*syn.* **P. auricomus**)
Similar to *P. variegatus*, but with a rich golden yellow variegation that is even more pronounced. The leaves are up to 18cm (7in) long, the culms an attractive purplish-green. The plant does well in shade or in pots.

H: to 1.5m (5ft), **S**: 4m (12ft)
❋❋❋ ◊ ◑ ☼ ☀ ♉

Sasa palmata f. nebulosa

Large-leaved, tropical-looking but fairly short bamboos, *Sasa* species run invasively and are unsuited to small gardens. *S. palmata* f. *nebulosa*, which has brown patterning on older culms, is the most vigorous. Plant it only in large containers or raised beds.

H: to 3m (10ft), **S**: indefinite
❄❄❄ ◊ ◖ ☼ ☀

Sasa veitchii

Smaller and better behaved than *S. palmata*, *S. veitchii* is probably the slowest spreader in the genus. In winter its leaf margins bleach to the colour of parchment. Unless you have plenty of space, site it with care – or plant it in a container.

H: to 1.5m (5ft), **S**: indefinite
❄❄❄ ◊ ◖ ☼ ☀

Semiarundinaria fastuosa var. viridis

This tallish bamboo is classed as a runner but behaves like a clumper. It makes an excellent screen or hedge, with dense plumes of long, narrow leaves, bright green culms, and a tough constitution.

H: to 12m (40ft), **S**: 4m (12ft)
❄❄❄ ◊ ◖ ☼ ☀

Bamboos (Shibataea–Yushania)

Shibataea kumasaca

One of the best small bamboos, *S. kumasaca,* from Japan, is clump-forming, slow-spreading, and dense. It forms a mound of broad, pointed leaves usually only 24in (60cm) high. It is ideal for small gardens, pots, and planting in gravel, but needs acid soil.

H: to 1.7m (6ft), **S**: 2m (6ft)
❋❋❋ ◊ ◗ ☼ ☀

Thamnocalamus crassinodus '*Kew Beauty*'

Among the most beautiful of all bamboos, *Thamnocalamus,* from the Himalayas, have many branches and small leaves. The culms of 'Kew Beauty' emerge powder blue, turning green then deep red with age.

H: to 8m (25ft), **S**: 4m (12ft)
❋❋❋ ◊ ◗ ☼

Thamnocalamus spathiflorus

This clump-forming but relatively fast-growing bamboo is bigger than *T. crassinodus* and has larger leaves. It may top 10m (30ft) but is usually less than half this in a garden situation. The green culms age to an unusual pinkish-brown.

H: to 10m (30ft), **S**: 6m (20ft)
❋❋❋ ◊ ◗ ☼

Yushania maculata

Although it is technically a clump-former, *Y. maculata* may become a runner, but it is never as invasive as *Sasa* or *Pleioblastus*. Small-leaved, arching, and graceful, it makes a fine screen or hedge. It is not suitable for small gardens or containers.

H: to 3.5m (11ft), **S**: 4m (12ft)
❋❋❋ ◊ ◗ ☼ ☼

Grasses (Anemanthele–Briza)

Anemanthele lessoniana
Pheasant's tail grass, once classed as a species of *Stipa*, is still sometimes sold as such. From New Zealand, it is semi-evergreen with gold and orange autumn-winter colour. It forms a fine-leaved, arching, domed tussock, with feathery, rather insignificant flowers.

H: 1m (3ft), **S**: 1.2m (4ft)
❋❋/❋❋❋ ◊ ◑ ☼ ◑

Arundo donax
Among the most statuesque of all non-woody plants, giant, or Provencal, reed is in fact a true grass. Common throughout the Mediterranean, where it regularly reaches 5m (15ft) tall, it has thick stems and large leaves, giving it the look of a tropical bamboo.

H: 4m (15ft), **S**: 2m (6ft)
❋❋/❋❋❋ ◊ ◑ ☼ ◑

Arundo donax *var.* versicolor
While undeniably beautiful, variegated selections of the giant reed are far less vigorous than the plain green-leaved form. They are also less hardy, so may be best planted out for summer, then dug up and brought under glass for the winter in cold areas.

H: 2.2m (7ft), **S**: 2m (6ft)
❋❋ ◊ ◑ ☼ ◑

Briza maxima
Quaking grass is among the most beautiful of annual grasses. It is well worth raising from seed for its large, unusual, early flowers that look like rattlesnake tails and rustle in the slightest breeze. It likes full sun but grows in most soils, even wet.

H: 30cm (12in), **S**: 23cm (9in)
❋❋❋ ◊ ◑ ◑ ☼

Briza minor
Essentially a scaled-down version of *B. maxima*, this pretty annual grass is even more floriferous, the flowers forming an ever-shifting haze above the foliage. Both species can be cut back hard in late summer to produce a second crop of leaves and flowers.

H: 23cm (9in), **S**: 15cm (6in)
❋❋❋ ◊ ◑ ◑ ☼

Grasses (Calamagrostis–Cortaderia)

Calamagrostis x acutiflora 'Karl Foerster'

Named selections of this hybrid feather reed grass are among the best for naturalistic, prairie styles of planting. 'Karl Foerster' is an upright grower with arrow-like flowerheads. Its leaves and flowers withstand winter well.

H: 1.5m (5ft), **S**: 1m (3ft)
❋❋❋ ◊ ☼

Calamagrostis x acutiflora 'Overdam'

Variegated 'Overdam' is more arching than 'Karl Foerster', making dense clumps of white-striped leaves. Its flowers emerge purple-tinged and are produced in a single flush. Shear back to enjoy another flush in late summer.

H: 1m (3ft), **S**: 1.2m (4ft)
❋❋❋ ◊ ☼

Calamagrostis brachytricha

Korean feather reed grass is useful because it will tolerate some shade, and its flowers – strongly tinted red when they emerge – appear late, usually in early autumn. Less upright than other feather reed grasses, it grows better on moist soils.

H: 1.2m (4ft), **S**: 1m (3ft)
❋❋❋ ◊ ◑ ☼ ◐

Calamagrostis emodensis

A pretty little grass recently introduced from Nepal, *C. emodensis* is barely half the height of most other species and cultivars, forming low clumps up to 1m (3ft) across made up of broad, arching leaves. It produces large, pink-tinged, *Miscanthus*-like flowers.

H: 60cm (24in), **S**: 1m (3ft)
❄❄❄ ◊ ◖ ☼ ☀

Chasmanthium latifolium

Native to woodlands and woodland edges, northern sea oats is a useful grass for shady areas, forming an arching mound of broad leaves and locket-shaped, *Briza*-like flowers. It is deciduous, but holds its structure well through winter. Dislikes drought.

H: 1.2m (4ft), **S**: 1m (3ft)
❄❄❄ ◖ ☀

Chionochloa conspicua

Probably the best known and hardiest of the New Zealand snow grasses (but not suited to very cold areas), plumed tussock grass is evergreen. It forms low, wide, arching mounds of pale green foliage. Large creamy seedheads hang from arching stems all summer.

H: 1.8m (6ft), **S**: 1.2m (4ft)
❄❄/❄❄❄ ◊ ◖ ☼

Cortaderia richardii

A relative of pampas grass, which it rather resembles, *C. richardii* has a mound of narrow evergreen leaves, producing arching shaggy plumes of white or cream flowers. The leaves can die back after flowering, and it is not reliably hardy in very cold areas.

H: 3m (10ft), **S**: 1.7m (5.5ft)
❄❄/❄❄❄ ◊ ◖ ☼ ♔

Grasses (Cortaderia–Deschampsia)

Cortaderia selloana

Pampas grass, planted to excess in the 1970s, is only now becoming desirable again. Few grasses are as large or so architectural. Very hardy and drought-tolerant, it makes an excellent hedge or screen. It is also ideal for the back of a border or with other tall grasses.

H: 3m (10ft), **S**: 2.5m (8ft)
❄❄❄ ◊ ◖ ☼

Cortaderia selloana '*Aureolineata*'

A "dwarf" selection still capable of topping 1.5m (5ft), 'Aureolineata' has leaves with broad, golden edging that becomes more pronounced and richly coloured as the season progresses. It is a good choice for smaller gardens, mixing well with other grasses.

H: 1.5m (5ft), **S**: 1.5m (5ft)
❄❄❄ ◊ ◖ ☼ ♔

Cortaderia selloana '*Pumila*'

The best plain green dwarf cultivar, 'Pumila' rarely reaches 2m (6ft), usually rather less. More free-flowering and hardier than its larger relatives, it has stout stems and long-lasting flowers. Less vigorous, too, it also mixes better with other plants.

H: 2m (6ft), **S**: 2m (6ft)
❄❄❄ ◊ ◖ ☼ ♔

Cortaderia selloana '*Silver Comet*'
A relatively new cultivar with wide white variegation to its leaves, 'Silver Comet' appears silver from a distance and is arguably the best variegated pampas grass. The plant has short flower stems, so plumes may look untidy among the leaves.

H: 2m (6ft), **S**: 2m (6ft)
❀❀❀ ◊ ◐ ☼

Cortaderia selloana '*Sunningdale Silver*'
Considered one of the best of the larger pampas cultivars, 'Sunningdale Silver' can top 3m (10ft), with large, pure white (or "silver") flower plumes that last well held on stout stems. It needs space to reach its full potential.

H: 3m (10ft), **S**: 2.5m (8ft)
❀❀❀ ◊ ◐ ☼ ♥

Deschampsia cespitosa '*Goldschleier*'
Hair grasses produce finely divided airy flowerheads above clumps of narrow leaves. A useful shade tolerant plant, German-bred 'Goldschleier' has golden yellow flowers. It is often sold as 'Golden Veil'.

H: 1.2m (4ft), **S**: 1m (3ft)
❀❀❀ ◐ ☼ ☀

Deschampsia cespitosa '*Goldtau*'
Another cultivar of tufted hair grass, 'Goldtau' is often sold as 'Golden Dew'. It is similar to, but slightly smaller than, 'Goldschleier', and grows well in informal plantings and gravel gardens, tolerating full sun if given enough moisture.

H: 1m (3ft), **S**: 75cm (30in)
❀❀❀ ◐ ☼ ☀

Deschampsia flexuosa '*Tatra Gold*'
This yellow-foliaged form of wavy hair grass is smaller than *D. cespitosa* and native to woodlands, so it copes with shady, moist conditions better than most true grasses. It produces reddish-brown flowers above greenish-gold leaves.

H: 15cm (6in), **S**: 15cm (6in)
❀❀❀ ◐ ☼ ☀

Grasses (Elymus–Hakonechloa)

Elymus magellanicus
Probably the most intensely blue of all ornamental grasses, this deciduous wheat grass from the tip of South America is not reliably hardy in colder areas. It forms a lax, sprawling clump with flowers that have a herringbone structure, similar to ears of wheat.

H: 45cm (18in), **S**: 45cm (18in)
❄❄/❄❄❄ ◊ ◖ ☼

Eragrostis curvula
African love grass has narrow leaves and a profusion of diffuse, feathery flowers. It is similar in structure to hair grass but not as hardy. Its narrow greyish-green leaves bleach to buff in autumn and, along with the flower stems, last well into winter.

H: 1m (3ft), **S**: 1.2m (4ft)
❄❄ ◊ ◖ ☼

Festuca glauca
Fescues are small, narrow-leaved, tussock-forming, largely evergreen grasses common in upland and moorland areas. Blue fescue's leaf colour ranges from greyish-green to intense, silvery blue. Small flowers emerge blue and age to brown.

H: 30cm (12in), **S**: 60cm (24in)
❄❄❄ ◊ ◖ ☼ ☀

Festuca glauca 'Elijah Blue'
Among the bluest and largest of the blue fescues, 'Elijah Blue' looks amazing when contrasted with larger yellow- and bronze-leaved grasses and sedges. It is free-flowering but in common with most fescues, it dislikes chalky and alkaline soils.

H: 30cm (12in), **S**: 60cm (24in)
❄❄❄ ◊ ◖ ☼ ☀

Festuca valesiaca 'Silbersee'
Raised in Germany, this grass is often sold under the name 'Silver Sea', although it is much more blue than silver. Similar to blue fescue but a semi-dwarf form that makes a neat cushion, this is a superb grass for containers or rock gardens.

H: 13cm (5in), **S**: 13cm (5in)
❄❄❄ ◊ ◖ ☼ ☀

Glyceria maxima *var.* variegata
Striped mana grass is broad-leaved, white-striped, and one of the best variegated grasses. It copes with moist to downright wet soils – growing well near water – and in semi-shade. However, it spreads vigorously and so can prove invasive.

H: 60cm (24in), **S**: 1m (3ft) or more
❀❀❀ ◗ ◖ ☼ ☼

Hakonechloa macra '*Alboaurea*'
Golden Hakone grass is an elegant, low, arching, slow-growing deciduous grass from Japan, with rich golden yellow leaves thinly striped with green and occasional splashes of white variegation. In autumn it flushes pink, then red-purple.

H: 25cm (10in), **S**: 1m (3ft)
❀❀❀ ◗ ☼ ☼ ♀

Hakonechloa macra '*Aureola*'
Very similar to 'Alboaurea', lacking only the white splashes, but possibly even more intensely gold-striped. In autumn the leaves turn a vivid purple. Both plants look magnificent in tall containers, which show off their domed forms to perfection.

H: 25cm (10in), **S**: 1m (3ft)
❀❀❀ ◗ ☼ ☼ ♀

Grasses (Helictotrichon–Miscanthus)

Helictotrichon sempervirens

Blue oat grass is evergreen and relatively small. More upright and with narrower leaves than *Elymus* (*see p.134*), it forms clumps up to 60cm (24in) high and has tall, spiky, oat-like flowers. It prefers full sun and is drought-tolerant.

H: 1m (3ft), **S**: 1m (3ft)
❄❄❄ ◊ ◐ ☼ ♈

Holcus mollis '*Albovariegatus*'

A beautiful variegated form of Yorkshire fog, or creeping soft grass, a species that is widely distributed across the north of Britain. It can make good ground cover and mixes well with small sedges. This grass may scorch brown in full sun.

H: 15cm (6in), **S**: 2m (6ft)
❄❄❄ ◊ ◐ ◑ ☀

Imperata cylindrica '*Rubra*'

One of the finest foliage plants bar none, Japanese blood grass's upright leaves emerge red-tipped, the redness spreading down the leaf like a stain in summer and peaking in autumn. In colder areas it is best grown in a pot and given winter protection.

H: 45cm (18in), **S**: 1.8m (6ft)
❄❄ ◊ ☼ ◑

Milium effusum '*Aureum*'

Perhaps the best yellow-foliaged grass, Bowles' golden grass prefers moist soils, but colours most brightly in full sun. Semi-evergreen, the leaves are broad and gently arching. Cut back after flowering to encourage a new crop of bright yellow leaves.

H: 30cm (12in), **S**: 30cm (12in)
❄❄❄ ◊ ◐ ☼ ◑ ♈

Miscanthus sinensis 'Gracillimus'

Eulalia grasses are large, upright, drought-tolerant, elegant, and much in vogue for prairie and gravel planting. This cultivar is a parent of many newer introductions, and is fairly typical, with narrow leaves and hanging, finger-like flowers.

H: 2m (6ft), **S**: 2m (6ft)
❋❋❋ ◊ ◊ ☼

Miscanthus sinensis 'Grosse Fontäne'

Also known as 'Big Fountain', this has broad leaves with silver midribs. Its late-summer flowers emerge silver flushed with red and open out into thin-fingered, silver, weeping plumes not unlike those of pampas grass.

H: 2m (6ft), **S**: 2m (6ft)
❋❋❋ ◊ ◊ ☼ ♔

Miscanthus sinensis 'Kleine Fontäne'

The name means 'Little Fountain' but this grass is usually only 30cm (12in) shorter than 'Grosse Fontäne'. Its long-lasting flowers often give structure all winter and only need cutting back the following spring.

H: 1.5m (5ft), **S**: 2m (6ft)
❋❋❋ ◊ ◊ ☼ ♔

Miscanthus sinensis 'Little Kitten'

A truly dwarf cultivar but just as free-flowering as its larger relatives, 'Little Kitten' adapts well to containers. It is the best choice for smaller gardens, since it mixes better with other grasses than its more vigorous cousins.

H: 30cm (12in), **S**: 30cm (12in)
❋❋❋ ◊ ◊ ☼

Grasses (Miscanthus)

Miscanthus sinensis *'Malepartus'*
This medium-sized eulalia grass is a good all-round choice for medium to large gardens. It has large, erect, reddish-brown flower plumes and is particularly attractive in autumn, when the foliage turns orange and russet before bleaching to buff.

H: 1.5m (5ft), **S**: 2m (6ft)
❋❋❋ ◊ ◗ ☼

Miscanthus sinensis *'Morning Light'*
A variegated cultivar with an elegant, fountain-like shape, wider at the top than the bottom. The narrow leaves have a wide central white stripe and yellow margins. Very late flowering, it may not flower at all in colder areas.

H: 1.5m (5ft); **S**: 1m (3ft)
❋❋❋ ◊ ◗ ☼ ♈

Miscanthus sinensis *'Rotfuchs'*
The name translates as 'Red Fox', aptly describing the reddish-brown flowers when they emerge in late summer and early autumn. Upright and forming a vertical clump, 'Rotfuchs' is a good choice for small to medium-sized gardens.

H: 1.2m (4ft), **S**: 1.5m (5ft)
❋❋❋ ◊ ◗ ☼

Miscanthus sinensis '*Silberfeder*'
One of the largest-flowering of the *Miscanthus*, 'Silberfeder', or 'Silver Feather', may produce flower stems up to 2.5m (8ft) tall. The foliage is shorter, each leaf with a prominent white midrib. Long-fingered flowers turn into silver, feathery seedheads.

H: up to 2.5m (8ft), **S**: 2m (6ft)
❄❄❄ ◊ ◐ ☼ ♈

Miscanthus sinensis '*Silberspinne*'
The name translates as 'Silver Spider', which could describe the flowers or leaves, since both are held out stiffly at 90 degrees from the stems, unlike most miscanthus which have an upright or arching habit. The long, lax flowers emerge red and fade to silver.

H: 1.2m (4ft), **S**: 1.2m (4ft)
❄❄❄ ◊ ◐ ☼

Miscanthus sinensis '*Strictus*'
'Strictus' is stiffly erect in habit with regular creamy-white bands on the foliage from midsummer, quite unlike the "normal" variegation that runs down the leaves. It is primarily grown for its dramatic foliage; flowers may appear up to late autumn.

H: 2m (6ft), **S**: 75cm (30in)
❄❄❄ ◊ ◐ ☼ ♈

Miscanthus sinensis '*Undine*'
A medium-sized cultivar that blooms from late summer. The flowers emerge an attractive copper-pink, ageing to buff-brown. It has good orange autumn colour and suits smaller gardens. The leaves and flowerheads stand up well to winter storms.

H: 1.3m (4½ft), **S**: 1.7m (5½ft)
❄❄❄ ◊ ◐ ☼ ♈

Grasses (Miscanthus–Panicum)

Miscanthus sinensis '*Variegatus*'
This medium-sized miscanthus has broad white stripes running down the lengths of the leaves. Forming an arching clump, it flowers reasonably well in early autumn, but the blooms do not last as well as those of most other cultivars.

H: 1.5m (5ft), **S**: 1m (3ft)
❀❀❀ ◊ ◗ ☼ ♈

Miscanthus sinensis '*Yakushima Dwarf*'
Confusingly, this is also the name of a dwarf cultivar so similar to 'Little Kitten' (*see p.137*) that some taxonomists regard them as the same plant. True 'Yakushima Dwarf' is bigger, with oatmeal-coloured flower plumes.

H: 1.2m (4ft), **S**: 1.2m (4ft)
❀❀❀ ◊ ◗ ☼

Miscanthus sinensis '*Zebrinus*'
The original horizontally banded *Miscanthus* forms more arching, slightly taller mounds of foliage than 'Stricta'. The banding is temperature-dependent, so don't despair if your zebra grass is plain green in spring – stripes should appear in midsummer.

H: 2.5m (8ft), **S**: 1.2m (4ft)
❀❀❀ ◊ ◗ ☼ ♈

Molinia caerulea *subsp.* caerulea *'Edith Dudszus'*

This cultivar of purple moor grass is a deciduous species native to much of Europe that obligingly sheds its dead leaves. It has fine leaves and purple flowers in summer, lasting well into autumn.

H: 60cm (24in), **S**: 60cm (24in)
❁❁❁ ◊ ◑ ☼

Molinia caerulea *subsp.* caerulea *'Variegata'*

An attractively variegated form of purple moor grass with golden yellow, irregular banding to its leaves. Half the height of 'Edith Dudszus', it forms low clumps and produces taller flowers in a paler shade of purple.

H: 30cm (12in), **S**: 60cm (24in)
❁❁❁ ◊ ◑ ☼ ♈

Panicum capillare

Witch grass is an attractive annual species with broad, strap-shaped leaves. It forms narrow, arching clumps, above which rise spider-like, greenish-yellow flowers. The seeds are edible and it has been used in its native US to make flour.

H: 60cm (24in), **S**: 30cm (12in)
❁❁❁ ◊ ◑ ☼ ☼

Panicum virgatum *'Dallas Blues'*

Switch grasses belong to a large genus of annuals and perennials, only a few of which are in cultivation. *P. virgatum* is a deciduous perennial prairie grass with fine, "midge-like" flowers above wide-leaved clumps. 'Dallas Blues' is free-flowering with blue-grey leaves.

H: 1.2m (4ft), **S**: 75cm (30in)
❁❁❁ ◊ ☼

Panicum virgatum *'Heavy Metal'*

An exceptionally upright selection of switch grass with leaves that are a greyer shade of blue, 'Heavy Metal' has typically fine, airy, purple-tinged flowerheads. All *P. virgatum* cultivars prefer full sun and good drainage, and cope well with chalky soils.

H: 1m (3ft), **S**: 1m (3ft)
❁❁❁ ◊ ☼

Grasses (Pennisetum–Spartina)

Pennisetum alopecuroides

Fountain, or foxtail, grasses are among the most beautiful of deciduous ornamental grasses. *P. alopecuroides* forms a low, mounded, fine-leaved clump, above which long stems of pinkish-brown, hairy, fox-tail-like flowers rise in late summer.

H: 75cm (30in), **S**: 1m (3ft)
❄❄❄ ◊ ◑ ☀

Pennisetum alopecuroides 'Hameln'

Semi-dwarf 'Hameln' is ideal for containers, and also useful as edging or at the front of borders. As with all *Pennisetum*, the compact flowerheads form a real contrast to other grasses with more diffuse flowers.

H: 45cm (18in), **S**: 80cm (30in)
❄❄❄ ◊ ◑ ☀

Pennisetum orientale

Oriental fountain grass is not as reliably hardy as *P. alopecuroides*. Although perennial, if short-lived, it is sometimes grown as an annual, forming sprawling tussocks of thin, fresh green leaves topped by long, narrow, pinkish-white flowerheads.

H: 1m (3ft), **S**: 1m (3ft)
❄❄ ◊ ◑ ☀ ⚱

Pennisetum orientale 'Karley Rose'

This form of oriental fountain grass has particularly long, elegant pink flowers that are produced from late summer. It is not the hardiest of grasses, which makes it unsuitable for colder areas.

H: 90cm (3ft), **S**: 90cm (3ft)
❄❄ ○ ◐ ☼ ☀

Pennisetum setaceum 'Purpureum'

This spectacular ornamental fountain grass is all purple, from its broad leaves to its large flowerheads. Opinions differ as to whether it is an annual or a short-lived perennial, but it is unable to survive even a few degrees of frost.

H: to 1.5m (5ft), **S**: to 1.2m (4ft)
❄ ○ ◐ ☼ ☀

Pennisetum villosum

Ethiopian fountain grass is on the tender side, though it should survive outdoors in sheltered spots and milder areas. It forms sprawling, bright green tussocks topped by short, wide, furry white flowers throughout summer and autumn.

H: 60cm (24in), **S**: 1m (3ft)
❄❄ ○ ◐ ☼ ♔

Phalaris arundinacea var. picta

Known as ribbon grass or gardener's garters, this deciduous broad-leaved grass has long been a favourite in cottage-garden borders. Its wide arching leaves are strongly striped with white, and very ornamental, though the flowers are inconspicuous.

H: 1m (3ft), **S**: 2m (6.5ft)
❄❄❄ ○ ◐ ☼ ☀

Spartina pectinata 'Aureomarginata'

A variegated form of North American prairie cord grass forming arching clumps of narrow, dark green leaves edged in gold, and comb-like, purple-tinged flowers. It likes damp to wet, but not permanently waterlogged, soil.

H: 1.2m (4ft), **S**: 1m (3ft)
❄❄❄ ◐ ● ☼ ☀

Grasses (Stipa)

Stipa calamagrostis
This feather grass forms a rounded mound of narrow, arching, deciduous leaves. Its soft, feathery, greenish-white flower plumes are so dense that they may obscure the foliage. They age to an attractive beige and persist most of the winter.

H: 1m (3ft), **S**: 1m (3ft)
✻✻✻ ◊ ◓ ☼ ☀

Stipa gigantea
Spanish oat grass is unmistakeable, with dark evergreen mounds of foliage above which rise huge stems of large, loose, golden, oat-like flowers. These appear in late spring and persist through summer well into winter, forming a semi-transparent screen.

H: 2m (6ft), **S**: 2m (6ft)
✻✻✻ ◊ ☼ ♆

Stipa tenuifolia
Often confused with *S. tennuissima* which it closely resembles, *S. tenuifolia* is smaller, producing broad mounds of fine, jade-green leaves topped with fine, airy, beige flowers from late spring to late summer. Like other stipas, it grows well on dry soils.

H: 45m (18in), **S**: 45cm (18in)
✻✻✻ ◊ ◓ ☼ ☀

Stipa tenuissima
A deservedly popular species with fine green leaves smothered in wispy flowerheads that look rather like a shaving brush. Emerging pink and ageing to white, they may obscure the foliage until late autumn, when its deciduous structure usually collapses.

H: 75cm (30in), **S**: 60cm (24in)
✻✻✻ ◊ ◓ ☼ ☀

Sedges and other grass-like plants (Acorus)

Acorus calamus

Sweet rushes, or sweet flags, are undemanding evergreen little plants in the arum family, thriving in moist to wet soils and shade. *Acorus calamus* forms stiff, tight, arching clumps of relatively broad leaves from a fleshy rootstock.

H: 45cm (18in), **S**: 45cm (18in)
❅❅❅ ◊ ♦ ☼ ☀

Acorus calamus '*Variegatus*'

A variegated selection with half green and half creamy-white leaves, often with an attractive pink tinge in colder weather. It holds its variegation well in shade and, like all *Acorus*, never becomes invasive. Grows well in pots.

H: 45cm (18in), **S**: 45cm (18in)
❅❅❅ ◊ ♦ ☼ ☀

Acorus gramineus

The smaller of the two *Acorus* in general cultivation, this Japanese native has thinner, more grass-like leaves and a less upright habit. It can form low-growing, evergreen carpets of ground cover, and spreads slowly by underground stems.

H: 20cm (8in), **S**: 30cm (12in)
❅❅❅ ◊ ♦ ☼ ☀

Acorus gramineus '*Hakuro-nishiki*'

A beautiful golden-leaved dwarf form selected and named in Japan that grows to barely half the height of the plain green species. 'Hakuro-nishiki' can bring a real splash of colour to shady areas and grows well in containers.

H: 10cm (4in), **S**: 20cm (8in)
❅❅❅ ◊ ♦ ☼ ☀

Acorus gramineus '*Ogon*'

One of the best variegated selections of sweet rush, the leaves are broadly and irregularly striped with cream to yellow variegation on a dark olive-green background. Almost as vigorous as the plain green species, it keeps its colouring even in full shade.

H: 25cm (10in), **S**: 25cm (10in)
❅❅❅ ◊ ♦ ☼ ☀

Sedges and other grass-like plants (Carex)

Carex buchananii
This is one of a group of bronze-brown to rust-red narrow-leaved sedges from New Zealand that contrast well with other coloured-leaved grasses and sedges. *Carex buchananii* forms upright clumps and its leaves often curl at the tips.

H: 60cm (24in), **S**: 60cm (24in)
❄❄/❄❄❄ ◊ ◐ ☼ ☀ ♈

Carex comans
Another bronze-leaved evergreen New Zealand sedge, *C. comans* and its selections have extremely narrow, thread-like leaves and a low-growing, sprawling habit. They look extremely attractive cascading over the edges of a tall pot.

H: 30cm (12in), **S**: 75cm (30in)
❄❄/❄❄❄ ◊ ◐ ☼ ☀

Carex elata 'Aurea'
One of the best known and most widely grown of all sedges, *C. elata* 'Aurea' is deciduous, shedding its leaves in autumn. It forms an upright, arching fountain of golden-yellow leaves thinly edged in green, preferring moist soils and part shade.

H: 75cm (30in), **S**: 1m (3ft)
❄❄❄ ◊ ◐ ☼ ☀ ♈

Carex flagellifera 'Coca-Cola'
An unusual dark brown-leaved cultivar of a particularly lax, low-growing New Zealand sedge, whose leaves sprawl in all directions from a central clump, often bleaching lighter towards their tips. Makes a good contrast to blue and yellow-leaved companions.

H: 45cm (18in), **S**: 1m (3ft)
❄❄/❄❄❄ ◌ ◗ ☼ ☀

Carex grayi
An evergreen from North America with broad green leaves and an upright habit, the mace sedge is named for its unusual fruit, which are covered in large spikes like mini medieval maces. It prefers moist soils and can be grown in shallow water.

H: 50cm (20in), **S**: 45cm (18in)
❄❄❄ ◌ ◗ ☼ ☀

Carex oshimensis 'Evergold'
This lovely little evergreen variegated sedge from Japan has broad golden-yellow bands down each leaf and only thin stripes of green. It has a low, arching habit and is attractive as ground cover or in containers.

H: 50cm (20in), **S**: 45cm (18in)
❄❄❄ ◌ ◗ ☼ ☀ ♈

Carex pendula
A British native with its own subtle appeal; unusually for a sedge it is grown primarily for its flowers. These are like brown catkins, are several centimetres long, and hang from arching stalks well above a dense clump of broad, fresh green leaves.

H: 1m (3ft), **S**: 1m (3ft)
❄❄❄ ◌ ◗ ☼ ☀

Sedges and other grass-like plants (Carex–Cyperus)

Carex pendula 'Moonraker'

Recently introduced, 'Moonraker' is a gorgeous variegated drooping sedge. New leaves emerge almost white in spring, developing rich creamy-yellow stripes and becoming greener as summer develops. It flowers well for a variegated plant.

H: 75cm (30in), **S**: 1m (3ft)
❀❀❀ ◊ ◆ ☼ ☀

Carex siderosticha 'Variegata'

Creeping broad-leaved sedge is a low-growing little deciduous sedge from Japan that spreads slowly by underground stems (rhizomes). 'Variegata' has narrow white stripes to the leaf margins, pink-tinged in cold springs. Great for ground cover.

H: 20cm (8in), **S**: 1m (39in)
❀❀❀ ◊ ◆ ☼ ☀

Carex tenuiculmis

Another New Zealand bronze sedge, *C. tenuiculmis* tends to be pale brown rather than bronze, particularly when grown in full sun. It forms a dense, upright tussock, the leaves arching towards their tips. Needs winter protection in colder areas.

H: 45cm (18in), **S**: 45cm (18in)
❀❀/❀❀❀ ◊ ◆ ☼ ☀

Carex testacea

The narrow leaves of *C. testacea* are an unusual olive-yellow with bronze highlights, quite unlike the colouring of other New Zealand species. Grow it in full sun, where it will turn a striking orange. However, it is not the hardiest of the New Zealand sedges.

H: 45cm (18in), **S**: 1m (3ft)
❀❀/❀❀❀ ◊ ◆ ☼ ☀

Carex trifida

An imposing evergreen sedge for moist areas and pond edges, *Carex trifida* forms large, bold clumps of fresh green leaves with an unusual "pleated" outline. The flowers are large and quite striking for a sedge, like mini bulrushes.

H: 1m (3ft), **S**: 1m (3ft)
❀❀❀ ◊ ◆ ☼ ☀

Cyperus eragrostis

Pale galingale is a native of tropical America. Although it is hardier than most *Cyperus*, it has the typical whorled, "umbrella-like" foliage surrounding unusual brown, wispy flowers. Like other *Cyperus*, it grows best with its feet in water.

H: 1m (3ft), **S**: 75cm (30in)
❄❄/❄❄❄ ◐ ◑ ☼ ☀

Cyperus involucratus

Umbrella sedge, or umbrella palm, is a handsome plant from Africa, with long, thick, leafless stems, each topped with "spokes" of bracts (not true leaves) surrounding the airy flowerheads. Sadly, it is not at all hardy and best grown indoors.

H: 1m (39in), **S**: 1m (39in)
❄ ◑ ☼ ☀ ▽

Cyperus papyrus

This magnificent water plant gave the pharaohs their paper, and grows to 5m (15ft) in the wild. Its stout stems are topped with wide, thread-like flowerheads. Unless you have a conservatory with a pool, it is better to grow one of the dwarf forms indoors.

H: 4m (12ft), **S**: 4m (12ft)
❄ ◑ ☼ ☀ ▽

Sedges and other grass-like plants (Equisetum–Ophiopogon)

Equisetum hyemale
Horsetails have been around for more than 300 million years. Rough horsetail, or scouring rush, is evergreen and produces ramrod-straight stems striped brown at the nodes. It is best confined to a pot to control its spreading tendencies.

H: 2.2m (7ft); **S**: 2m (6ft) or more
❄❄❄ ◐ ● ☼ ◑

Equisetum ramosissimum *var. japonicum*
Smaller than rough horsetail, this evergreen variety is from Japan but is fully hardy. Although shorter, the stems are thicker than those of rough horsetail, with striped brown and near white nodes. Confine it to a pot.

H: 1m (3ft); **S**: 2m (6ft) or more
❄❄❄ ◐ ● ☼ ◑

Juncus effusus *f.* spiralis
Arguably more a curiosity than a true ornamental, this is the largest of the so-called spiral rushes in cultivation. Evergreen, it has unusual dark green corkscrew-like stems – a mutation that seems to be common in several rush species.

H: 30cm (12in); **S**: 60cm (24in)
❄❄❄ ● ☼ ◑

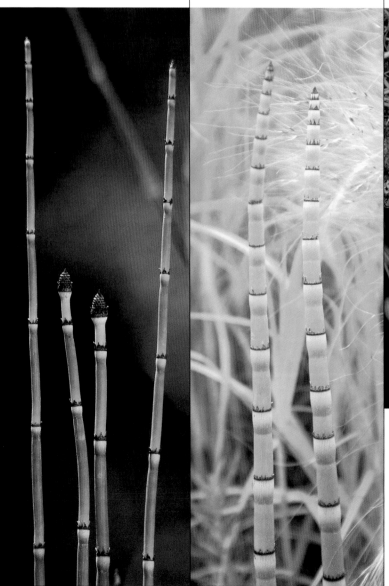

Juncus patens 'Carman's Gray'
A named form of California grey rush, 'Carman's Gray' has a distinct blue-grey colouring, particularly when grown in full sun, which it prefers. Typically upright but free-flowering for a rush, though the small brown flowers are of little ornamental merit.

H: 30cm (12in); **S**: 60cm (24in)
❄❄❄ ● ☼ ◑

Luzula nivea

Woodrushes are some of the best plants for dry, heavy shade, such as under trees. Snowy woodrush, from southern Europe, is evergreen, clump-forming, and spreads by rhizomes (underground stems), though not to excess. It has attractive white flowers.

H: 45cm (18in); **S**: 60cm (24in)
❆❆❆ ◊ ◊ ☼ ☀

Luzula sylvatica 'Aurea'

This yellow-foliaged selection of the British native greater woodrush is one of the best plants for introducing a splash of colour into shady areas. It grows well on virtually any soil, is drought-tolerant, and evergreen in all but the coldest areas.

H: 30cm (12in); **S**: 60cm (24in)
❆❆❆ ◊ ◊ ☼ ☀

Ophiopogon jaburan 'Vittatus'

The so-called "mondo grasses" have narrow, strap-like, evergreen leaves but are members of the *Convolvulus* family, as the flowers indicate. The leaves of 'Vittatus' are irregularly striped with creamy-white to yellow stripes. Low-growing and elegant.

H: 15cm (6in); **S**: 30cm (12in)
❆❆❆ ◊ ◊ ☼ ☼

Sedges and other grass-like plants (Ophiopogon–Uncinia)

Ophiopogon japonicus
This species has the narrowest leaves in the genus, which are an attractive blue-green in full sun, making it look perhaps the most grass-like. A very useful, drought-tolerant little plant that makes an excellent alternative lawn that never needs cutting.

H: 15cm (6in); **S**: 30cm (12in)
❉❉❉ ◊ ◐ ☼ ☀

Ophiopogon planiscapus 'Nigrescens'
This is one of the few plants with truly black leaves. Evergreen, it spreads slowly by runners and produces small white flowers in summer, which are followed by shiny, jet-black berries.

H: 15cm (6in); **S**: 30cm (12in)
❉❉❉ ◊ ◐ ☼ ☀ ♆

Phragmites australis *subsp.* australis *var.* striatopictus 'Variegatus'
Norfolk reed is an aggressive colonizer. This variegated selection with broad creamy-white striping to the leaves is much less invasive but still best confined to a basket in small ponds.

H: 1.4m (4.5ft); **S**: 2m (6ft)
❉❉❉ ◊ ● ☼ ☀

Schoenoplectus lacustris *subsp.* tabernaemontani *'Zebrinus'*

Schoenoplectus are true bullrushes, a common name usually used for *Typha*, the reed maces. Deciduous 'Zebrinus' is less invasive than the plain green species and has the same horizontal banding as some *Miscanthus* cultivars.

H: 1.5m (5ft); **S**: 1.2m (4ft)
❄❄❄ ◐ ◆ ☼ ◑

Typha latifolia

Reed mace, or cats' tails, form fat, brown, cigar-shaped flowers up to 30cm (12in) long and are out-and-out moisture lovers, growing best with their roots in water. *T. latifolia* is a good deciduous water plant for large ornamental ponds and lakes only.

H: 3m (10ft); **S**: 2m (6ft)
❄❄❄ ◐ ◆ ☼ ◑

Typha minima

Much better suited to smaller gardens, *T. minima* reaches barely 60cm (2ft) tall, with narrow, grass-like foliage bearing small brown flowerheads that are more round than cigar-shaped. Best confined to a basket in small ponds to control its spread.

H: 60cm (2ft); **S**: 1.2m (4ft)
❄❄❄ ◐ ◆ ☼ ◑

Uncinia rubra

Hook sedges are handsome rust-red true sedges, similar in appearance to some of the bronze *Carex* species and, like them, from New Zealand. However, *Uncinia* prefer damp to wet soils and partial shade, making them useful for bog and water gardens.

H: 30cm (12in); **S**: 1m (3ft)
❄❄❄ ◐ ◆ ☼ ◑

Suppliers

Bamboos

Amulree Exotics
The Turnpike, Norwich Road
Fundenhall
Norfolk NR16 1EL
Tel/Fax: 01508 488101
Email: SDG@exotica.fsbusiness.co.uk
www.turn-it-tropical.co.uk

Architectural Plants
Nuthurst
Horsham
West Sussex RH13 6LH
Tel: 01403 891772
Fax: 01403 891056
Email: enquiries@architecturalplants.
  com
www.architecturalplants.com

Architectural Plants (Chichester)
Lidsey Road Nursery
Woodgate, Nr Chichester
West Sussex PO20 3SU
Tel: 01243 545008
Email: chichester@architecturalplants.
  com
www.architecturalplants.com

Drysdale Garden Exotics
Bowerwood Road
Fordingbridge
Hampshire SP6 1BN
Tel: 01425 653010
(National Collection of Bamboos)

Fulbrooke Nursery
Home Farm
Westley Waterless
Newmarket
Suffolk CB8 0RG
Tel/Fax: 01638 507124
Email: bamboo@fulbrooke.co.uk
www.fulbrooke.co.uk
(Bamboos and grasses; open by
appointment only)

Moor Monkton Nurseries
Moor Monkton
Nr York
Yorkshire YO26 8JJ
Tel/Fax: 01904 738770
Email: sales@bamboo-uk.co.uk
www.bamboo-uk.co.uk

Norfolk Bamboo Company
Vine Cottage, The Drift
Ingoldisthorpe
King's Lynn
Norfolk PE31 6NW
Tel: 01485 543935
Email: Lewdyer@hotmail.com
www.norfolkbamboo.co.uk

Pan Global Plants
The Walled Garden
Frampton Court
Frampton-on-Severn
Gloucestershire GL2 7EX
Tel: 01452 741641
Fax: 01453 768858
Email: info@panglobalplants.com
www.panglobalplants.com

PW Plants
'Sunnyside' Heath Road
Kenninghall
Norfolk NR16 2DS
Tel/Fax: 01953 888212
Email: pw@hardybamboo.com
www.hardybamboo.com
(Many species of bamboo planted out
in a large demonstration garden)

Whitelea Nursery
Whitelea Lane
Tansley, Matlock
Derbyshire DE4 5FL
Tel: 01629 55010
Email: whitelea@nursery-stock.
freeserve.co.uk
www.uk-bamboos.co.uk
(Open by appointment only)

Grasses, sedges, and grass-like plants

The Alpine and Grass Nursery
Northgate
West Pinchbeck
Spalding
Lincolnshire PE11 3TB
Tel: 01775 640935
www.alpinesandgrasses.co.uk
(Open by appointment only)

The Big Grass Co.
Hookhill Plantation
Woolfardisworthy East
Nr Crediton
Devon EX17 4RX
Tel: 01363 866146
Email: alison@big-grass.com
www.big-grass.com
(Mail order only; open by
appointment only)

Bramley Lodge Garden Nursery
Beech Tree Lane
Ipplepen
Newton Abbott
Devon TQ12 5TW
Tel: 01803 813265

Bridge Nursery
Tomlow Road
Napton
Nr Rugby
Warwickshire CV47 8HX
Tel: 01926 812737
Email: pmartino@beeb.net
www.Bridge-Nursery.co.uk

**Fordmouth Croft Ornamental
Grass Nursery**
Fordmouth Croft
Meikle Wartle
Inverurie
Aberdeenshire AB51 5BE
Tel: 01467 671519
Email: Ann-Marie@fmcornamental
 grasses.co.uk
www.fmcornamentalgrasses.co.uk
(Mail order; open strictly by
appointment only)

Hall Farm Nursery
Vicarage Lane
Kinnersley
Nr Oswestry
Shropshire SY10 8DH
Tel/Fax: 01691 682135
Email: hallfarmnursery@ukonline.
 co.uk
www.hallfarmnursery.co.uk
(Grasses and water plants)

Highdown Nursery
New Hall Lane
Small Dole
Nr Henfield
West Sussex BN5 9YH
Tel/Fax: 01273 492976
Email: herbs@btinternet.com
(Herbs and grasses specialists)

Knoll Gardens
Hampreston
Wimborne
Dorset BH21 7ND
Tel: 01202 873931
Email: enquiries@knollgardens.co.uk
www.knollgardens.co.uk
(Nursery also has large demonstration
gardens of grass-based plantings)

The Plantsman's Preference
Office:
Hopton Road,
Garboldisham, Diss
Norfolk IP22 2QN
Tel: 01953 681439
Fax: 01953 688194

Nursery:
Church Road
South Lopham
Nr Diss
Norfolk IP22 2LW
Tel: 07799 855559
Email: tim@plantpref.co.uk
www.plantpref.co.uk

Pukka Plants
Count House Farm
Treglisson
Wheal Alfred Road
Hayle
Cornwall TR27 5JT
Tel/Fax: 01872 271129
Email: sktrevena@gmail.com
www.pukkaplants.co.uk

Wallace Plants
Lewes Road Nursery
Lewes Road
Laughton
East Sussex BN8 6BN
Tel: 01323 811729
Email: sjk@wallaceplants.fsnet.co.uk
www.wallaceplants.fsnet.co.uk

Meadow seed mixtures

Chiltern Seeds
Bortree Stile
Ulverston
Cumbria LA12 7PB
Tel: 01229 581137
Fax: 01229 584549
Email: info@chilternseeds.co.uk
www.chilternseeds.co.uk

Emorsgate Wild Seeds
Limes Farm
Tilney All Saints
King's Lynn
Norfolk PE34 4RT
Tel: 01553 829028
Fax: 01553 829803
Email: enquiries@emorsgate-seeds.
 co.uk
www.wildseed.co.uk

Farm Direct
2 Dovers Yard
Flimby Brow
Flimby
Maryport
Cumbria CA15 8SX
Tel: 01900 819923
www.farmdirectonline.co.uk

Landlife Wildflowers
National Wildflower Centre
Court Hey Park
Liverpool L16 3NA
Tel: 0151 7371819
Fax: 0151 7371820
www.wildflower.org.uk
(Also has a series of demonstration
plots open to the public)

Pictorial Meadows
Manor Lodge
115 Manor Lane
Sheffield S2 1UH
Tel: 0114 2762828
Fax: 0114 2706271
Email: info@pictorialmeadows.co.uk
www.pictorialmeadows.co.uk
(A range of free-flowering mixtures
of native and exotic wildflowers)

Index

Index

Acknowledgements

The publisher would like to thank the following for their kind permission to reproduce their photographs:

(Key: a-above; b-below/bottom; c-centre; l-left; r-right; t-top)

2 Airedale: Sarah Cuttle. **4** DK Images: Peter Anderson (c). Paul Whittaker/PW Plants (b). **6–7** Airedale: David Murphy: Designer: Tom Stuart Smith, Daily Telegraph Garden, Chelsea Flower Show 2006. **8** Airedale: Amanda Jensen (bl). **8–9** DK Images: Steve Wooster. **9** Sarah Cuttle (b). Airedale: Amanda Jensen (t). **10** DK Images: Brian T North/Designer: John Brookes (tr); Roger Smith (tl). Airedale: Amanda Jensen: Designer: Tom Stuart Smith, Daily Telegraph Garden, Chelsea Flower Show 2006 (b). **12** Dianna Jazwinski (b). Airedale: David Murphy: Designer: Stuart Perry, Halifax Garden, Chelsea Flower Show 2006 (tr). **13** The Garden Collection: Jonathan Buckley/Designer: Mark Brown. **14** Mike Newton (bl). **14–15** Mike Newton. **16** Airedale: Amanda Jensen: Knoll Gardens. **17** Airedale: Amanda Jensen (bl); Amanda Jensen: Knoll Gardens (t); David Murphy (br). **18** Airedale: Amanda Jensen: Designer: Ruth Marshall, Mencap & Cater Allen Bank Garden (Seeing the Whole Picture), Chelsea Flower Show 2006 (b); Amanda Jensen: Designers: Marcus Barnett & Philip Nixon, The Savills Garden, Chelsea Flower Show 2006 (t). **19** The Garden Collection: Marie O'Hara/Scenic Blue Design Team. **20–21** DK Images: Steve Wooster. **21** Leigh Clapp: Copse Lodge (b). Airedale: Amanda Jensen: RHS Wisley (t). **22** Airedale: Amanda Jensen (l); David Murphy (br). **24–25** Airedale: Amanda Jensen: Designer: Geoff Whiten, The Pavestone Garden (Garden of Tranquillity), Chelsea Flower Show 2006. **25** Sarah Cuttle: Designer: Ian Rochead, The City Workers Retreat, Chelsea Flower Show 2006 (b). Harpur Garden Library: Marcus Harpur:Designer: Justin Greer for Holly Oldcorn (tr). **26** Dianna Jazwinski. **27** Amanda Jensen (br); David Murphy: Designer: Tom Stuart Smith, Daily Telegraph Garden, Chelsea Flower Show 2006 (t). **28–29** Airedale: David Murphy: Designer: Sarah Eberle in collaboration

with Andrew Herring, Bradstone Garden, Chelsea Flower Show 2006 (b); David Murphy (t). **29** John Glover: Hillcroft Road, Walsall (tr). **32** Mike Newton. **37** Airedale: David Murphy (bc). **38** DK Images: Peter Anderson (cl) (bl) (br); Mark Winwood (tr). **39** Andrew Lawson: Designers: Biddy Bunzl and James Fraser. **42** DK Images: Steve Wooster/Designer: Geoff Whiten (r). Mike Newton (l). **43** Airedale: Amanda Jensen: Designer: Emma Dawson, Capel Manor College, Time Line, Hampton Court Flower Show 2006 (l). DK Images: Peter Anderson/Designer: Pat Garland (r). **52** Airedale: David Murphy. **58–59** Airedale: Amanda Jensen: Designer: Trevor Tooth Garden Practice (Love, Life & Regeneration), Hampton Court Flower Show 2006. **61** The Garden Collection: Jonathan Buckley: Design: David and Mavis Seeney, Upper Mill Cottage, Kent. **62** DK Images: Peter Anderson (bl). Airedale: Amanda Jensen (tl) (cl). Crocus.co.uk (tr). **63** Airedale: Amanda Jensen: Designers: Marcus Barnett & Philip Nixon, The Savills Garden, Chelsea Flower Show 2006. **64** Garden World Images: Tony Cooper (br). **65** Garden World Images: Tony Cooper. **66** Paul Whittaker/PW Plants (bl). **67** Paul Whittaker/PW Plants. **69** Mark Bolton: Designers: Alison & Scott Evans/The Big Grass Company, Devon. **70** The Garden Collection: Gary Rogers (cr). Crocus.co.uk (tr). **71** The Garden Collection: Gary Rogers: Designers: Alex Daley & Alice Devaney, Tatton Park 2005. **75** Mark Bolton: Pinsla Garden, Cornwall. **76** Airedale: Amanda Jensen (tl). **77** Airedale: David Murphy. **78** Airedale: Amanda Jensen, Knoll Gardens, Dorset (tl). **79** Airedale: Amanda Jensen, Knoll Gardens, Dorset. **81** Clive Nichols: RHS Garden, Wisley. **82** DK Images: Mark Winwood (bl). **83** Airedale: Sarah Cuttle. **84** Crocus.co.uk (br). **87** Airedale: Amanda Jensen: Designer: Tom Stuart Smith, Daily Telegraph Garden, Chelsea Flower Show 2006. **88** Crocus.co.uk (bl). **89** John Glover: Designer: Geoff Whiten, Hampton Court Flower Show 1995. **92** Sarah Cuttle (bl). **93** Sarah Cuttle: Designers: Nigel Boardman & Stephen Gelly, Boardman, Gelly & Co (The Urban Jungle), Hampton Court Flower Show 2006. **94** Airedale:

Amanda Jensen (bl). **95** Airedale: Amanda Jensen: Designer: Graham Pockett: The Zoological Society of London (Gorilla Garden), Chelsea Flower Show 2006. **96** Airedale: Amanda Jensen (cr) (bl) (br). John Carter (tl). **97** Airedale: Amanda Jensen: Designer: Graham Pockett: The Zoological Society of London (Gorilla Garden), Chelsea Flower Show 2006. **99** DK Images: Peter Anderson: Chelsea Flower Show 2001. **118** Sarah Cuttle (r). **122** DK Images: Peter Anderson (tl). **123** Bamboo Garden (tr). Paul Whittaker/PW Plants (br). **124** Crocus.co.uk (c). **125** Paul Whittaker/PW Plants (tr). **126** Paul Whittaker/PW Plants (tl). **128** Paul Whittaker/PW Plants (tc) (bc). **129** Airedale: Amanda Jensen (tc). **134** Paul Whittaker/PW Plants (br). Crocus.co.uk (cb). **139** Crocus.co.uk (tr). **141** Crocus.co.uk (tc). **144** Crocus.co.uk (br). **146** DK Images: Mark Winwood (bl). **147** Crocus.co.uk (br). **148** Airedale: Amanda Jensen (bl) (bc). Paul Whittaker/PW Plants (tl). **150** Airedale: David Murphy (c). Paul Whittaker/PW Plants (br). **152** Crocus.co.uk (tc).

All other images © Dorling Kindersley. For further information see: www.dkimages.com

Every effort has been made to trace the copyright holders. We apologise in advance for any unintentional omission and would be pleased to insert the appropriate acknowledgements in any subsequent edition.

Dorling Kindersley would also like to thank the following:
Editorial assistance: Fiona Wild, Mandy Lebentz; *Index*: Michèle Clarke

Airedale Publishing would like to thank the following:
Crocus.co.uk; Knoll Gardens; Paul Whittaker at PW Plants; Thompson & Morgan (Tel: 01473 695200).